Progress in Precision Agriculture

Series editor
Margaret A. Oliver, Soil Research Centre, University of Reading, Reading, Berkshire, United Kingdom

This book series aims to provide a coherent framework to cover the multidisciplinary subject of Precision Agriculture (PA), including technological, agronomic, economic and sustainability issues of this subject. The target audience is varied and will be aimed at many groups working within PA including agricultural design engineers, agricultural economists, sensor specialists and agricultural statisticians. All volumes will be peer reviewed by an international advisory board.

More information about this series at http://www.springer.com/series/13782

Søren Marcus Pedersen • Kim Martin Lind
Editors

Precision Agriculture: Technology and Economic Perspectives

Springer

Editors
Søren Marcus Pedersen
Department of Food and Resource Economics
University of Copenhagen
Frederiksberg C, Denmark

Kim Martin Lind
Department of Food and Resource Economics
University of Copenhagen
Frederiksberg C, Denmark

ISSN 2511-2260 ISSN 2511-2279 (electronic)
Progress in Precision Agriculture
ISBN 978-3-319-68713-1 ISBN 978-3-319-68715-5 (eBook)
https://doi.org/10.1007/978-3-319-68715-5

Library of Congress Control Number: 2017958136

Printed on acid-free paper

This Springer imprint is published by Springer Nature
The registered company is Springer International Publishing AG
The registered company address is: Gewerbestrasse 11, 6330 Cham, Switzerland

Preface

This is the first book in the new book series on Progress in Precision Agriculture established by Springer. The book series was originally proposed because there is currently no such series devoted to this topic and there are few books devoted to specialist topics in precision agriculture (PA). The one on *Geostatistical Applications for Precision Agriculture* has been successful and is listed as one of Springer's most downloaded books. Therefore, this overall background led to the concept of this book series. In addition, precision agriculture is becoming increasingly important with issues of population growth, decreasing availability of adequate arable land and land degradation that has resulted from the intensive use of land following the Green Revolution.

The series will provide a framework for a wide range of subject areas and provide a 'state-of-the-art' view of where the research frontier lies. Topics will include the economic, technological, agronomic and sustainability issues of PA. The books will aim to provide a teaching resource for university and college course leaders and a basis for new researchers to develop research concepts and for farm and environmental managers, civil servants and so on.

The first book in the series, *Precision Agriculture: Technology and Economic Perspectives,* was selected because there is a lack of material on the economics of precision agriculture. This is very closely linked to the technology in PA, which involves major investment by farmers. The book covers a wide range of topics and will provide a sound basis for many groups in PA such as farmers, advisors, policy makers, engineers, economists and students that have an interest in precision agriculture.

Chapter 1 gives an introduction to precision agriculture (PA) with a short historical review of the development and the status of current available technologies. Part of this description also provides an overview of some of the economic barriers and technical obstacles when applying variable-rate application. The aim of the chapter is to provide readers with a foundation for what is to come in the subsequent chapters.

Chapter 2 provides an overview of the current smart farming technologies that are available among farmers or are about to become available in the near future.

Smart farming technologies (SFT) cover a range of different aspects of precision agriculture including data acquisition technologies, data analysis and evaluation technologies and precision application technologies. Furthermore, the economic impact that each SFT has in comparison to conventional agricultural practices is assessed.

Chapter 3 gives a first assessment of the potential economic impact of variable-rate fertilizer application. It shows that precision farming technologies that aim to identify the economically optimal input rate often fail to provide considerable economic advantages for the farmer. This phenomenon can be explained by flat payoff functions, which are relevant for many agricultural production processes.

Chapter 4 describes the relevance and profitability of different precision herbicide application technologies; two weed detection technologies and a low-dose decision-support system (DSS) are analysed.

Chapter 5 focuses on the economics and perspectives of site-specific irrigation management with a focus on automated furrow irrigation, which is a new technology being developed commercially and offered to farmers in Australia. This analysis considers the economics of adopting an improved precision irrigation technology from two different perspectives: economic and environmental.

Chapter 6 provides an analysis of the feasibility of auto-steering systems and controlled traffic farming systems (CTF). In this chapter, four different machinery scenarios were tested in four fields each, and the main objective was to compare two different route planning systems under economic criteria apart from the best operational route coverage design criterion. The results show the considerable potential of advanced route planning designs and further optimization of farming systems.

Chapter 7 focuses on the profitability of controlled traffic in grass silage production. From a farmer's perspective, the potential profitability of converting to CTF is determined by the existing machinery system and the required investment for a CTF conversion. Moreover, the on-farm profitability potential will be determined by the site-specific conditions in terms of yield response from CTF, opportunities to produce other cash crops and the knowledge and the involvement required in setting up and maintaining the CTF system.

Chapter 8 examines pre-commercial and autonomous systems that could be implemented in the near future, namely, robotic crop cultivation systems. The potential economic benefits from the use of agricultural robots under specific conditions and constraints are described.

Chapter 9 provides a general description of farm management information systems (FMIS). The state of the art is presented depicting the new functionalities included in FMIS and how they can connect the farm to the external context and stakeholders. The authors delve into the functionality of FMIS to understand how precision agriculture can improve the allocation of costs to the final product by managing PA in a better way.

Chapter 10 focuses on how precision farming as a sustainable technology can help to increase yields and reduce the environmental impact of crop farming; a case study from Estonia is presented in which PF is one of the technologies among others.

Chapter 11 provides an overview of various studies of the adoption of precision agriculture. It shows that PA affects the performance of farms positively, even though its benefits vary according to the size of farms as well as their location.

Chapter 12 closes the book by examining the perspectives of precision farming in a broader policy context. Agriculture is faced with contrasting opposing requirements from society at large. Precision agriculture can be part of the response to often conflicting issues by the use of technologies in a precise and targeted approach that reduce resource use and increase yield. Furthermore, the growing demand for higher-value food products in terms of health benefits and quality requires traceability and information about production processes and resource use. These correspond with the possibilities offered by precision agriculture technology. The general movement towards greater integration in food supply chains is a natural extension of the requirements for traceability and product information, which are integral parts of precision agriculture.

Frederiksberg, Denmark — Søren Marcus Pedersen
Frederiksberg, Denmark — Kim Martin Lind
Reading, UK — Margaret A. Oliver
August 2017

Contents

Contributors

Hans Alvemar SOYL Sverige AB, Skara, Sweden

Hans Andersson Swedish University of Agricultural Sciences, Uppsala, Sweden

Jelena Ariva Estonian University of Life Sciences, Institute of Economics and Social Sciences, Tartu, Estonia

Raiko Aste Estonian University of Life Sciences, Institute of Economics and Social Sciences, Tartu, Estonia

Athanasios T. Balafoutis Agricultural University of Athens, Athens, Greece

Institute of Bio-economy & Agro-technology Centre of Research & Technology Hellas, Volos, Greece

Bert Beck Institute for Agricultural, Fisheries and Food Research (ILVO), Merelbeke, Belgium

B. Simon Blackmore Harper Adams University, Newport, Shropshire, UK

Dionysis Bochtis Department of Engineering, Faculty of Science and Technology, Aarhus University, Aarhus, Denmark

Centre for Research and Technology-Hellas (CERTH), Institute for Research and Technology – Thessaly (IRETETH), Volos, Greece

Giacomo Carli Department of Strategy and Marketing, The Open University Business School, Milton Keynes, UK

Robert Farquharson The University of Melbourne, Melbourne, VIC, Australia

Spyros Fountas Agricultural University of Athens, Athens, Greece

Markus Gandorfer Chair of Economics of Horticulture and Landscaping, Technical University of Munich, Freising, Germany

Manuel Gómez-Barbero European Commission – Joint Research Centre (JRC), Directorate D – Sustainable Resources – Economics of Agriculture, Seville, Spain

Peter Kryger Jensen Department of Agroecology, Aarhus University, Aarhus, Denmark

Per Kudsk Department of Agroecology, Aarhus University, Aarhus, Denmark

Kim Martin Lind Department of Food and Resource Economics, Faculty of Science, University of Copenhagen, Frederiksberg C, Denmark

Andreas Meyer-Aurich Leibniz-Institute for Agricultural Engineering and Bioeconomy e.V, Potsdam, Germany

Jens Erik Ørum Department of Food and Resource Economics, University of Copenhagen, Copenhagen, Denmark

Hans Henrik Pedersen Aarhus University, Aarhus, Denmark

Søren Marcus Pedersen Department of Food and Resource Economics, Faculty of Science, University of Copenhagen, Frederiksberg C, Denmark

Erika Pignatti Check Fruit-NSF Italy, Bologna, Italy

Efthymios Rodias Department of Engineering, Faculty of Science and Technology, Aarhus University, Aarhus, Denmark

Claus G. Sørensen Department of Engineering, Faculty of Science and Technology, Aarhus University, Aarhus, Denmark

I. Soto-Embodas European Commission – Joint Research Centre (JRC), Directorate D – Sustainable Resources – Economics of Agriculture, Seville, Spain

Maria Rita Tagliaventi Department of Management, University of Bologna, Bologna, Italy

Zisis Tsiropoulos Agricultural University of Athens, Athens, Greece

Tamme van der Wal Wageningen Environmental Research (Alterra), Wageningen, The Netherlands

Frits K. Van Evert Wageningen University & Research, Wageningen, The Netherlands

Jürgen Vangeyte Institute for Agricultural, Fisheries and Food Research (ILVO), Merelbeke, Belgium

Rando Värnik Estonian University of Life Sciences, Institute of Economics and Social Sciences, Tartu, Estonia

Jon Welsh Cotton Research and Development Corporation, Narrabri, Australia

Vilma Xhakollari Department of Agricultural Sciences, University of Bologna, Bologna, Italy

Chapter 1
Precision Agriculture – From Mapping to Site-Specific Application

S.M. Pedersen and K.M. Lind

Abstract This chapter gives an introduction to Precision Agriculture (PA) with a short historic pathway of the development and the status of current available technologies. Part of this description also provides an overview of some of the economic barriers and technical obstacles when applying variable-rate application. This chapter also notes that the adoption of several specific variable-rate application technologies have been modest in recent years. However, in contrast the adoption of auto-steering has been significant in the last decade. The last section describes the overall aim of the book and an overview of each chapter in the book. Each chapter address a different topic starting with an overview of technologies that are currently available, followed by specific Variable-Rate Technologies such as VRT fertilizer application, VRT pesticide application, site-specific irrigation management, Auto-steering and Controlled Traffic Systems. Finally, the chapter looks into new developments of autonomous systems with an example of robotic seeding, farm information management in precision agriculture and different methods on the adoption of PA. The last chapter focuses on how PA can fulfil the current policy trends on environmental regulations.

Keywords Precision agriculture • Adoption • History • Economics

1.1 Introduction

The introduction of tractors and combine harvesters has changed many farms from small labour-intensive units where the individual farmer had detailed knowledge about each field to large farm holdings with large fields to monitor. In some countries, farms can reach several thousand hectares of arable land. This trend was

S.M. Pedersen (✉) • K.M. Lind
Department of Food and Resource Economics, Faculty of Science, University of Copenhagen, Frederiksberg C, Denmark
e-mail: marcus@ifro.ku.dk; kim@ifro.ku.dk

S.M. Pedersen, K.M. Lind (eds.), *Precision Agriculture: Technology and Economic Perspectives*, Progress in Precision Agriculture,
https://doi.org/10.1007/978-3-319-68715-5_1

further driven by the introduction of commercial fertilizers and chemical weed control practices that pushed the development towards even more labour-saving production processes and larger fields. The larger the individual farm and fields became, the less knowledge the farm manager had about each field.

The introduction of precision farming is to some extent a step back in time. On the one hand, PA is aimed at large holdings with a farm and capital structure that enables them to invest in expensive systems. On the other hand, it is a means to get farm management back to small scale farming processes with detailed knowledge about small units and management zones and enable farmers to treat each unit, whether it is a piece of land or an animal, with the same care as farmers did in previous times. This development is facilitated by the help of smart technologies that allow the farmer to gain detailed knowledge about the field and subsequently to treat the field accordingly.

> Precision farming can be defined as *the application of technologies and principles to manage spatial and temporal variability associated with all aspects of agricultural production for the purpose of improving crop performance and environmental quality* (Pierce and Nowak 1999).

Precision agriculture is described in the literature by various terms such as Precision farming (PF) , Site-Specific input Application (SSA), Site-Specific Agricultural Technology and Variable-Rate Treatment (VRT). In the following, these terms will be used synonymously. A broader term is Smart Agriculture that also appears to cover later technical developments such as auto steering systems, controlled traffic farming and autonomous systems like agricultural robots.

As a management tool, precision agriculture consists of four elements: geographical positioning (GPS), gathering information, decision support and variable-rate treatment. Yield mapping could be regarded as a fifth component, in which yield mapping enables the farmer to monitor the actual outcome from varying inputs (Pedersen 2003). However, yield mapping is also a tool that enables information to be gathered about previous years of yield on the field, which can be used as decision support for designing the next input strategy. Figure 1.1 illustrates the link between different technical systems and sensors in precision farming from geographical positioning and sensing systems to decision support, variable-rate input application and route planning.

Recently, several technologies that are related to precision farming have emerged from GPS systems, yield mapping, smart sensors and auto steering systems.

These systems should ideally enable the farmer to increase yield, save nutrients and replace labour time with efficient sensing and decision-support systems that can increase profitability on the farm and reduce the negative environmental impact. In that sense, precision farming follows other cost and labour saving innovations such as reduced tillage and GM crops – the recent developments of semi-autonomous systems and farm robots is likely to speed up that process further.

Precision farming has, since the beginning with the combined introduction of the first general purpose technology GPS (global positioning system) and GIS (geographical information system) with yield meters and maps in the late 1980s,

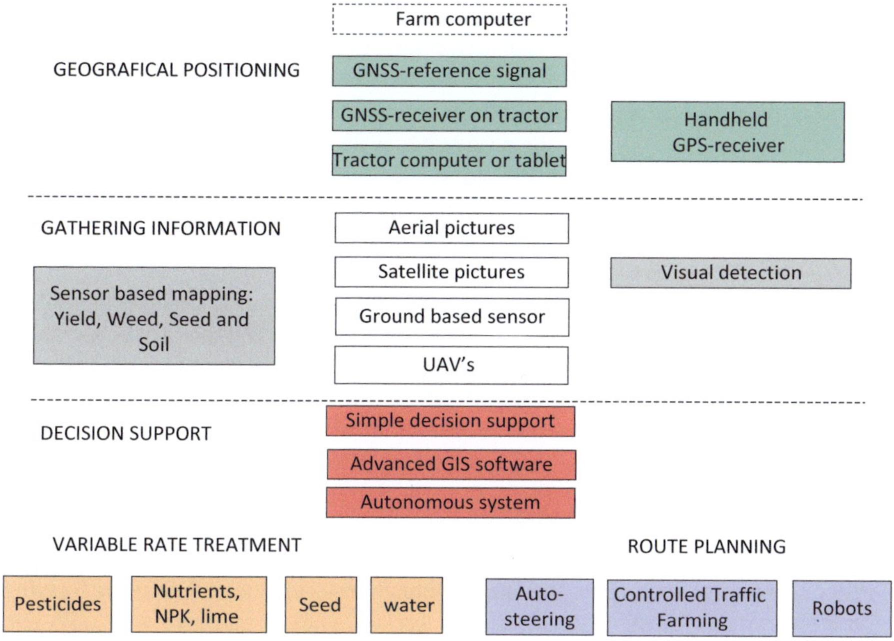

Fig. 1.1 Sensors and technical systems applied in precision agriculture

developed in many directions and the term PF has become quite complex covering many technical solutions.

It is always difficult to set the borders for a book like this as different elements of precision farming may be applied in different cropping systems and with different combinations of technology. There is, however, one core element that seems to be a part of all these individual technologies and that is the GPS or GNSS (global navigation satellite system) that enables site-specific treatment by the farmer.

In this study, we have decided to focus mainly on arable farming and we tend to disregard horticulture and orchards although many precision farming systems have emerged in these areas. In addition, we will mainly address the potential adoption of Precision Agriculture in a European context on farms with medium to large scale fields.

The focus of this book will be on a farm management perspective rather than society perspectives although some of the chapters will explore social values, adoption, environmental impact and sustainability issues. Basically, the book will address two main questions:

Which technologies are available for farmers and will some of these precision farming technologies be profitable for farmers to use? In addition, the book will cover aspects on farmers' adoption and perception of these technologies and the broader impact of PF as a technology to comply with current political trends, which focus on the environment.

With this book we will cover some potential PA technologies only and their economic potential. In the last decade hundreds of new start-up companies have been engaged in developing new sensors, apps and farm management systems that incorporate GNSS, positioning and GIS mapping systems in the agricultural sector. Many of these innovation attempts have reached a commercial level, whereas others are still at the development stage – but in any case a tremendous development has taken place around the world to develop better systems in America, Europe, Asia as well as targeted applications in developing countries. One factor is technology, however, an even more important factor is the deployment of technology.

The potential profitability of selected PA technologies will undoubtedly depend on many variables such as farm and field size, crop types, soil type, soil type, degree of specialization at the farm, on farm labour costs and access to finance and collateral for the individual farmer. In addition, the individual farmer may decide about adoption depending on his current investments in machinery and expected time of replacement. Other elements that may hinder or promote adoption is access to training and extension services related to precision farming technologies.

From a technical point of view, adoption may also be influenced by access to the internet, current speed of computers and on the current mix of farm equipment at the farm.

Since the mid 1980s, precision agriculture has developed from the use of the first GPS system and yield maps to the latest applications with controlled traffic farming with RTK systems and UAVs (Unmanned Aerial Vehicles) mounted with GPS and cameras for crop scouting. Meanwhile, the first attempt has been made to make autonomous or semi-autonomous systems that can carry out tasks to some extent on their own with no or little surveillance. Today, other navigation systems have emerged or are about to be launched in the near future, which might complement the GPS systems. Table 1.1 shows a brief historical pathway of precision farming since its early introduction in the mid 1980s.

Table 1.1 History of precision agriculture

1970–1980	Introduction of GPS as a general purpose technology
	First yield meter mounted on a combine harvester
1984	First yield maps introduced (with GPS)
1991	Application maps (GIS based) introduced and first attempt with variable-rate technology
1995–1998	Groundbased and satellite/aerial sensing systems to measure crop status (chlorophyll content)
1999–2002	Introduction of soil electrical conductivity measurements and aerial/satellite images to measure crop status
2000	Introduction of RTK systems applied in agriculture
2000–2002	First attempt with weed detection systems and precise seeding
2005	systems
2003	Introduction of auto-steering in agriculture
2008	Implementation of first controlled traffic systems among farmers
	Introduction of UAVs (drones) for application maps
2015	Introduction of first robotic systems in high value crops/horticulture

Parallel to this technical and commercial development, several scientific conferences have developed mainly in North America, Australia and Europe and later on in Asia and South America.

European Conference on Precision Agriculture (ECPA), and the International Conference on precision Agriculture in the USA, which is held every second year in Europe and the USA, respectively. They both cover topics on the economics and adoption of precision farming. Other conferences such as the Asian-Australian Conference on Precision Agriculture have likewise taken up topics on economics and farm management in precision farming. In addition, numerous other agricultural engineering conferences and workshops as well as conferences on agricultural economics and Farm Management have taken up issues on economics of PA.

These conferences have to some extent included specific topics on the economic viability of selected precision farming technologies and adoption of precision farming in different regions, especially topics on auto-guidance systems, variable-rate nitrogen and lime application have been numerous. Little attention has so far been on the economic feasibility of variable-rate pesticide application and variable-rate irrigation systems.

A number of books have focused on various forms of precision farming technologies with chapters on economic impacts like Handbook of Precision farming edited by Ancha Srinivasan from 2006 and Precision agriculture for sustainability and environmental protection, edited by Oliver, Bishop and Marchant from 2013, which also address topics on sustainability and economic perspectives. Nevertheless, little attention has been given to assess the broader economic perspectives for farmers adopting precision farming. This book will try to bridge this gap with a focus on the financial and economic viability and adoption of PF among farmers. The book is meant to target readers such as researchers, agricultural engineers, academics, farm advisors and graduate students within the area of crop production and farm management as well as students with an interest in precision farming and farm management.

This book is divided into 12 chapters in which some of the most recent developments and aspects of precision farming are addressed. Although the intention of this book is to give an overview of some of the most promising technologies within precision farming from an economic point of view, each chapter can be read individually if a reader wishes to focus on one particular topic.

Authors will present cases from different countries with a main focus on the perspectives of using precision farming in Europe. Since precision farming as a farm technology will benefit from scale advantages because of relatively large investment costs, the adoption in North America, Australia and other countries with large field areas are among the global front runners and early adopters.

1.2 Current PA Technologies

Precision farming is an information-intense and data intensive concept. Large datasets (Big Data) are required to generate maps as well as to display and make interpretations of specific variables. In this area, new technologies are under development or have been developed during the last decade. To assess the impact and profitability of precision farming, it is necessary to consider farming and farm management as a whole. In some regions, the use of manure and slurry from animals will have an impact on arable farming. In this case, it may be necessary to include the distribution and access of manure and slurry in the models to assess precision farming practices. Organic farming practices are also users of animal manure in a rotational cropping system and should therefore be similarly considered in relation to PA practices. In this book, the focus will be on conventional arable farming and little attention will be given specifically to organic farming practices.

1.3 Variable-Rate Application

The collection of data about spatial variation within the field is important, however, of more importance is how to use this information in variable-rate input application and, furthermore, to manage this input in the most economically efficient way.

If we consider a field with two different soil types A and B with 50% of the entire field each. Soil type A is a nutritional clay soil and soil type B is a sandy soil with a lower yield potential.

In conventional farming, inputs are traditionally distributed uniformly over the entire field with the same amount of input per unit on each site of the field. This method is convenient and requires no site-specific information about the soils.

In principle, precision farming is a matter of distributing inputs site-specifically to gain marginal yield benefits from the same field or to save inputs. With a completely homogenous field, the potential benefits from site-specific application is in principle zero, but with heterogeneous fields it should be possible to gain a benefit according to the variation in the yield potential or input saving potential from each field unit. However, even though each unit has its own yield potential, varying the application of nutrients may not necessarily imply an extra marginal economic benefit. It all depends on the yield slope of the yield response curves and the marginal net benefit of adding an extra input site-specifically, and to what extent the necessary site-specific information is available in order to make that decision.

Figure 1.2 illustrates yield response functions of one input for two soil types A and B. If we assume that the prices and benefits of these inputs and yields per unit are the same, regardless of the application level, then these S-shaped yield response functions may also be regarded as gross margin response functions.

The function is S-shaped to illustrate the diminishing return to increased use of inputs that is common among crop production functions. At some point, a particular

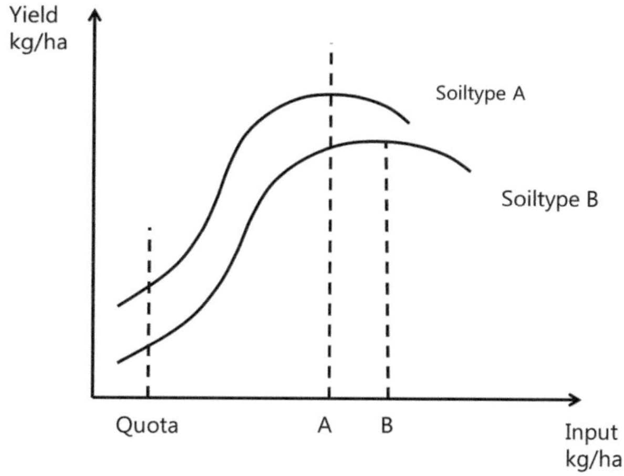

Fig. 1.2 Illustration of a single field with yield response functions of inputs for two soil types A and B

field may not gain extra yield and even get reduced yields and gross margins from adding one extra unit of input. One could imagine that by applying too much nitrogen to a field; at some point the crop yield will decrease as shown in Fig. 1.2. For soil type A, this peaking point is at input level A and for soil type B this input level could be at point B.

By using precision farming with site-specific application, the farmer may in principle be able to use information about the soil type or other site-specific information to make an optimising decision and to improve the distribution of the input.

However, it may also be the case that due to environmental restrictions or other restrictions on access to inputs, the farmer may be able to use only a certain amount of fertilizer as with a quota indicated by the dotted line. At this input level, the yield is higher for soil type A than soil type B.

Assuming that we have only a certain amount of nitrogen as indicated by the quota and dotted line. At this amount, the yield per kilogram of nitrogen is higher for soil type A than for soil type B. If we have a quota of 100 kg of nitrogen and divide it into 50 kg for Soil type A and 50 kg for soil type B then we will gain a higher yield from soil type A. The logical suggestion would then be to add more nitrogen to soil type A and less to soil type B because the yield is higher at soil type A. However, at the point of the quota (and just below and after this point) the "shapes" of the two curves are exactly the same. This means that if we take some of the nitrogen from soil B and give an extra amount to soil type A, then indeed we get more yield on soil type A, but also less yield on soil type B since we took nitrogen from this area. In fact, since the slopes of the two curves are exactly the same the yield loss from soil type B will be exactly the same as the yield increase at soil type A, meaning that the overall yield from the two fields from making that differentiated application will be exactly the same. Since the slope of the curve is the same for both soil types at that input level, it is therefore not beneficial to redistribute the quota amount of input because an additional increase in yield by using more inputs on one soil type will be offset equally with less yield on the other soil type.

This simple example illustrates the increasing and decreasing marginal benefits of inputs and is in our view a key management and economic issue that occurs within site-specific application.

Another issue that illustrates the complexity is that the farmer may not even know about the underlying yield response function if he does not have full information about the soils or he does not have sufficient information about how well the field is drained or how much it will rain in the days to come while he is distributing his fertilizer. He may not have full knowledge about the site-specific soil water content at the root zone at the right time. To make a good decision about how much fertilizer to use at a given location at a given time – all this information is needed. For many farmers, this information is not readily available and a precise decision-support system is not on offer. Therefore, the benefits realised from variable-rate fertilizer application have been shown to be modest until now.

In this example, the input was nitrogen, but in principle the farm manager may have to consider many other inputs like lime, pesticides, seeds and water in the same manner. What is the marginal net benefit from distributing inputs in the best possible way? In this example, we just used the soil type (either A or B) as the spatial variable, but as more knowledge has been gained about precision farming more information should be included in the decision-making process.

From a farm economic point of view, nitrogen should therefore be applied at a rate where the farmer's expected payoff is optimized. Optimal application rates will vary across fields with different soil types, but also across space and time because of the variation in crop growing conditions. Obtaining the full value of the economic benefits from site-specific applications requires an accurate indication of the variation in crop status in relation to yield potential within the field (Robertson et al. 2008) as well as an indication of the future weather conditions.

Crop status and crop growth status can be measured by using sensors that can be mounted on tractors, airborne UAVs or from satellite pictures. In all these systems, the spectral reflectance of the crop is made into a vegetation index and used as an indicator of the greenness and amount of crop biomass in the field (Evert et al. 2012).

Recently, it was proposed to adopt UAV mounted sensors to provide farmers with better information about the crop status and to enable site-specific application of fertilizers. This is another solution to existing on-board canopy sensing systems, like the Yara-sensor that is applied in Europe, which provide the possibility of online application. So far, the relatively high costs associated with these PF systems have limited the adoption. It is expected that new and lower cost systems may allow more farmers to take up the precision fertilization technology. The combined integration of crop modelling, UAVs and better decision-support systems may provide farmers with improved applications that include information about the spatial and temporal variation of crops in order to perform site-specific application of fertiliser.

Site-specific application is considered as a future eco-innovative technology by many agronomists. However, its adoption by farmers and in particular on farms with small field sizes is still relatively low. Some of the barriers are the complexity of the technology that consists of loose components, the lack of adequate service providers with 'knowhow', the initial investment costs of the systems compared with the yearly benefits and reliability (in particular) of the satellite based monitoring systems due to cloud cover, but similarly on tractor mounted systems.

Several studies have focused on farmers who already practice some kind of site-specific application or yield mapping and their perceptions on a general level. Factors that have been addressed of concern are, time consumption, accuracy of the system, compatibility with other systems and cost versus return (Pedersen and Pedersen 2002; Pedersen et al. 2003).

By using the latest crop monitoring systems (by combining hyperspectral cameras, UAVs, with crop simulation modelling) and integrating this into a comprehensive and directly deployable service may overcome these barriers.

Recently, the availability of site-specific variable-rate equipment (like sprayers and spreaders with GPS and on-board rate controllers) is increasing. Soil and crop simulation models can quantify yields and N-leaching under different soil, environmental and management conditions (Basso et al. 2011). Models can be useful for tactical management of N fertilizer rate in relation to water availability and radiation based on rainfall amounts). Basso et al. (2011, 2012) has demonstrated the economic and environmental advantages of using site-specific fertilizer applications (different zones within the field) and time (over different years) with the SALUS-model. So far, most of the focus has been on applying N fertilizer aimed at increasing crop yield. However, recently more focus has also been directed towards the environmental impact (nitrate leaching and nitrous oxide emissions) in relation to N fertilization strategies. By using a systems approach in crop and soil models, better management strategies are offered that both minimize the environmental impact and increase farmers' profitability. A number of models (including DAISY, FASSET and SALUS have been applied under a wide range of environmental and soil conditions for different crops (Basso et al. 2007, 2010, 2012; Pedersen 2003; Ritchie 1998; Senthilkumar et al. 2009).

1.3.1 Other Variable Inputs and Benefits

In principle, other inputs like potassium, phosphorus and lime can be applied site-specifically, and a conventional soil analysis that focuses on potassium, phosphorus and pH-values in the soil can be conducted site-specifically on grids and in management zones. These three inputs are rather "stable factors" and will be removed with the crop during harvest according to the crops' requirements. However, these inputs are also considered to have a relatively low economic impact. Lime is a low cost input that is applied every 5 years depending on the soil conditions and pH value, and potassium and phosphorus are usually applied in much smaller amounts compared with nitrogen. With lime application, – a low pH value means that the soil has a high acidity. In cases with low pH values, lime is added in order to balance the pH-value.

Studies on variable-rate lime application in the USA have shown that annual returns could increase by more than 20€ ha^{-1} based on simulation models for soybean and corn (Bongiovanni and Lowenberg-DeBoer 2000). A study by Wang et al. (2003) also indicated additional returns to variable rate lime application in combination with nitrogen application.

Site-specific control of pesticides are practices that have received increasing attention. A number of studies have investigated the economics of site-specific weed management (Swinton 2005; Franco et al. 2017). It was initially considered to have a large economic potential in many cropping systems because weeds are expected to be located in patches on the field (Leiva et al. 1997; Daberkow 1997; Audsley 1993; Gerhards et al. 1999; Christensen et al. 1997, 2014). However, since automated detection of weeds still needs improvements in terms of detection and low cost mapping, manual and visual weed detection is often the only practical solution at present. Timmermann et al. (2003) indicate that cost savings could range between 20 and 40 € ha^{-1} with site specific weed management in different arable crops. These costs savings are related to cost savings of herbicides, however, with GNSS it may also enable the farmer to obtain knowledge about his exact location during a specific spraying operation. For instance, it may be convenient or even labour saving to know the location in the field exactly if he suddenly has to end an operation at night or when refilling the sprayer. In that case, it is convenient to use the GNSS location coordinates and return back and continue next day without concerns about how to find the location. A study by Franco et al. (2017) indicated that the potential gains of site-specific herbicide application decrease significantly with increased precision in spraying. Research into insecticides has also gained interest recently. A study with aphids and ladybird beetles in cereals indicated that variable-rate spraying with sensor technology could reduce insecticide use by 13% on average (Dammer and Adamek 2012).

Site-specific seeding and in particular improved distribution of the seed in patterns instead of seeding in rows have also gained interest from researchers in precision farming. Studies have found that a better distribution of seeds can increase yields compared with traditional rows (Heeje 2013b).

Until recently, most remote sensing of crops in Precision Agriculture has been conducted with airoplanes or through satellite imaging to cover large areas in a short time. Recently, the Sentinel satellite images, with a resolution of 10 ×10 m per pixel, have become available to farmers every 5 days free of charge. These satellite images enable farmers to estimate N uptake from maps of Normalized Difference Vegetation Index (NDVI). Major disadvantages with these systems are the lack of flexibility and clouded conditions in many European regions, which makes it difficult to use these remote images. Other systems have been ground based and tractor mounted sensors, however, for these systems to be cost-efficient sensing operations must be carried out with other operations in the field like harvesting, spreading of fertilizers etc.

Recently, the development of UAVs (unmanned eerial vehicles) that can fly underneath the clouds has offered a new and flexible system, which provides images with high spatial resolution (1 pixel <10 cm). A light-weight hyperspectral camera can then be mounted on the UAV to make images to determine biomass index and density and site-specific crop growth. A UAV operator requires less training than with manned airoplanes and investment costs are lower.

So far, most farmers have not been able to process all the available farm information data in an optimal management system in order to make optimal decisions.

As illustrated above with site-specific application, the decision support needed to maximize net benefits is rather complicated and relies on many sources and management of these data that are both spatial and timely.

Big-data handling has become an integral part of precision farming together with the on-going use of crop and soil sensors and yield maps aimed at providing good decision support. However, so far the marginal revenues from site-specific application of nitrogen have been modest.

In parallel to site-specific application of inputs, guidance systems such as light bar systems and automatic guidance systems have been developed. In principle, light bar guidance and automatic guidance are the same except that with automatic guidance the driver does not have to steer after a light bar, but instead the vehicle steers automatically based on an automatic steering algorithm and GNSS signals that go directly to the electric and hydraulic steering actuators. A study by Heeje (2013a) showed that the area for the adoption of automatic guidance systems should be more than about 450 hectares in small grains in order to become profitable.

Auto-guidance is thereby a new way of utilising the GNSS system compared with site-specific application. Auto-guidance systems enable farm machinery to follow driving lines reducing overlaps of the tractor on the field. An auto-guidance system helps farmers to reduce fuel costs, input costs, time, labour costs, soil compaction and increases overall field efficiency as well as operator comfort and work quality. Auto-guidance systems can be used for many field operations such as seeding, tillage, planting, weeding and harvesting (Abidine et al. 2002). A further development of Auto-guidance systems that has proved economically viable on large farms is controlled traffic farming (CTF). The CTF enables farmers both to reduce overlap as with auto guidance and reduce soil compaction and additional fuel consumption on the field (Jensen et al. 2012).

The pathway and **development of new technologies** will often take different directions and speed depending on the complexity of the technology. S. Davies developed a model that characterises two different types of innovations: A and B. The first model is the A group, which illustrates the diffusion pattern for technologies that are fairly simple to use. The learning effects of this simple technology is initially very high but after some time the productivity gains from that technology will be limited implying that the curve falls and stabilises at a given level. The second group B of technologies is fairly complex to use (Fig. 1.3).

Variable-rate fertilizer application can be characterised as a fairly complex technology to use. It requires many systems to be integrated and time to be familiar with the technology and the productivity gains from that technology will take place at a much later stage compared with auto-steering for instance. The concept of variable-rate application is highly technical and requires skills that go beyond technical knowledge. It requires an understanding of the interrelated impact that various data and information have on yields, crop quality and nutrient leaching.

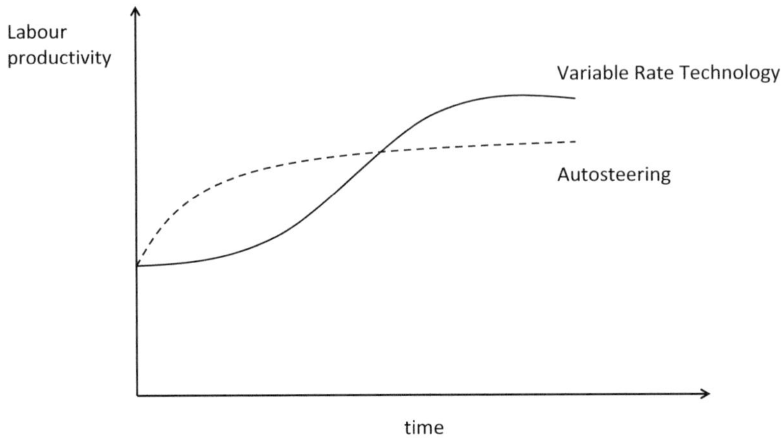

Fig. 1.3 Labour productivity for group A (autosteering systems) and group B Variable rate technology over time (Source: Based on Pedersen 2003; Davies 1979)

The adoption and diffusion of various techniques for practising precision farming depends therefore on the provision of advice, and technical as well as agronomic and biological guidelines.

In addition to the benefits from variable-rate application and reduced overlap and better route planning in the field, precision farming can also be a tool to trace the commodity from the field to the end consumer and thereby it can help to identify the geographical origin of final products in the supermarket. For some consumers, that information may be of value and they might even be willing to pay a premium for that information. In that case, supermarkets and supply companies may have an interest in providing farmers an incentive to use GPS on their farms.

1.4 Adoption of Precision Farming Systems

So far, there have been several surveys on the adoption and perception of site-specific application systems in Europe and North America (Reichardt and Jürgens 2008; Pedersen et al. 2001, 2003; Lawson et al. 2011).

Adoption of precision agriculture technologies is mainly driven by higher expected profits, OECD (2016). Consequently, the technologies that have demonstrated the highest earning potentials disseminate faster. Southern European countries are lagging behind in the adoption of new technologies in agriculture due to relatively small farms sizes (Lawson et al. 2011).

Table 1.2 shows the adoption of PA-technologies in England.

The table shows that positioning systems are the most adopted PA-technology, which is related to the documented significant higher expected profits from this technology. However, variable-rate application, although adopted less by farmers, is

Table 1.2 Farms using PA-technologies, % of farm holdings in England

PA-techologies	2009	2012
GPS	14	22
Soil mapping	14	20
Variable-rate application	13	16
Yield mapping	7	11
Telemetry	1	2

Source: OECD (2016)

still used by a significant proportion of farmers. Clearly, the table shows that positioning systems have had the highest growth rate from 2009 to 2012 relative to variable-rate application, which is most likely to be associated with the latter's mixed economic results.

A European market report from 2012 estimates that GNSS penetration into EU tractors will rise from 7.5% in 2012 to 35% in 2020 and currently about 70–80% of new farm equipment that is sold has some form of PF component inside. In addition, it is expected that average device GNSS/RTK prices will decrease by about 30% from 2012 to 2022, (EU Report 2014).

The United States Department of Agriculture (USDA) Economic Research Service has released a study on the adoption, use and portability of precision agriculture. In the study by Schimmelpfennig (2016), a survey on adoption shows that precision agriculture technologies were used on roughly 30–50% of maize/corn and soybean acres in the United States in the years 2010–2012.

This survey also allowed the examination of production and financial information for a large sample of farms. By focusing on three different technology systems, information mapping, guidance systems and variable-rate technology, the study shows that the use of GPS mapping had an impact on net returns of almost 2%. Guidance systems raise net returns by 1.5%. Variable-rate technology (VRT) raises net returns on maize farms by 1.1%. Yield mapping is mostly to be used on maize/corn and soybean crops although the use of yield maps has increased in other crops like wheat, peanuts and rice, Schimmelpfennig (2016).

A recent farm survey from 2016 from the State of Kansas in the US by Miller et al. (2017) with 348 farms shows that 228 farms (66% of the sample) have adopted auto-guidance systems on their farms. About 47% of the farms use automated section control, but only 17% of the farms use variable rate technology to apply seeds at site-specific rates. Variable-rate fertilizer application was used by 25% of the farmers.

In particular, the adoption of auto-steering systems has increased significantly in the last decade. A study by Lawson et al. (2011) shows a similar trend with high adoption of auto-guidance systems among large farms in Germany and Denmark. Compared with other technologies, the adoption of site-specific weed management has been slow in most countries, and farmers are often reluctant to implement PF technologies because of the high costs of investing in the technology and lack of technical compatibility with existing machinery (Pedersen 2003).

So far, PA have received most attention from countries in North America, Europe and Australia. However, recently more and more attention is coming from Asia and South America. About 90% of the 115.6 million of total farms in India has an area of less than 4 ha in size (Mondal and Basu 2009). A similar pattern with small farms can be observed in many other countries in Asia and Africa.

As investments in GPS systems and PA technologies are quite expensive and the financial burden should be distributed on a significant acreage before the returns from the field can cover the costs. The general perception is, therefore, that PA cannot be applied to small-scale farms in developing countries. To find the right technology for small scale farms to start with is therefore a challenge.

Some low cost technologies could be useful. The chlorophyll meter and leaf colour chart (LCC) are simple and portable tools that can be used for crop N status measurements in rice fields to determine the use of N application in real time (Mondal and Basu 2009).

However, so far most of the PF systems with GPS monitoring, auto-steering and VRT as the main technologies are designed for relatively large-scale farms located in mainly North America, Australia, Europe and South America.

1.5 The Aim and Organisation of this Book

The book is intended for readers who have an interest in precision farming and the economic potential for applying modern technology in arable farming. The book will provide a broad economic and technical insight into the area of advanced farming and cropping systems. In each area, it is the intention to provide the non-specialist with an update of some of the current precision farming technologies and their financial or economic potential among farmers. Each chapter address a different topic starting with an overview of the broad spectre of technologies that are currently available and subsequently followed by specific variable-rate technologies like, VRT fertilizer application, VRT pesticide application, site-specific irrigation management, auto-steering and controlled traffic systems. The following chapters look into new developments of autonomous systems with an example of robotic seeding, farm information management in precision farming and different methods on the adoption of PF. The last chapter focuses on how PF can fulfil the current policy trends on environmental regulations. As precision farming and farm information management have faced tremendous development in recent years and are expected to continue to develop further at high speed together with easier handling of big data sources, it is recommended that this book is updated within a relatively few years.

Chapter 2 of this book gives an overview of the current smart farming technologies that are available among farmers or are about to become available in the near future. Smart farming technologies (SFT) cover a range of different aspects of precision agriculture. This chapter provides an overview of data acquisition technologies, data analysis and evaluation technologies and precision application technologies. Furthermore, the reader can find a technical description of the technologies included in each category accompanied by a taxonomy of all SFT in terms of farming system type, cropping system, availability, level of investment and farmers' motives to adopt them. Finally, the economic impact that each SFT has in comparison to conventional agricultural practices is given.

Chapter 3 provides a first assessment of variable-rate treatment. It considers the potential economic impact of **variable-rate fertilizer application.** This chapter describes a framework for the economic assessment of site-specific fertilizer application and harvest management. It also provides an overview of selected studies and a future perspective of this technology. It concludes that precision farming technologies that aim to identify the economically optimal input rate (e.g. site-specific fertilizer application) often fail to provide considerable economic advantages for the farmer. This phenomenon can be explained by flat payoff functions, which are relevant for many agricultural production processes. Economically more promising technologies from a theoretical point of view are precision farming approaches that enable higher product prices by achieving specific product qualities (e.g. site-specific harvest management).

The purpose of precision farming technologies in relation to herbicide use is to reduce herbicide cost and to reduce environmental impact from spraying, but at the same time to control and maintain an acceptable weed population. Another purpose is to increase the spraying capacity, to reduce the number of sprayer refills, get faster treatment and a larger area per operation, and finally to minimize time-consuming activities like weed detection. Chapter 4 describes the relevance and profitability of four precision herbicide application technologies, two weed detection technologies and a low dose decision-support system (DSS) is analysed.

Chapter 5 focuses on the economics and perspectives of site-specific irrigation management with a focus on automated furrow irrigation, which is a new technology being developed commercially and offered to farmers in Australia. Improvements in water, fertilizer and labour efficiencies are possible with the more precise management of irrigation water, albeit with initial capital and ongoing management costs.

This analysis considers the economics of adopting an improved precision irrigation technology from two different perspectives: economics and environmental.

Chapter 6 provides an analysis of the feasibility of auto-steering systems and controlled traffic farming systems (CTF). In this chapter four different machinery cases were tested in four fields each and the main objective was to compare two different route planning systems under economic criteria apart from the best operational route coverage design criterion. The results show that there are significant reductions in the costs of operation which vary from 9–20%, depending on the specific machinery and field configurations. Such results show the considerable potential of advanced route planning designs and further optimization of farming systems.

Chapter 7 focuses in particular on the profitability of controlled traffic in grass silage production. From a farmer's perspective, the potential profitability of converting to CTF is determined by the existing machinery system and the required investment for a CTF conversion. Moreover, the on-farm profitability potential will be determined by the site-specific conditions in terms of yield response from CTF, opportunities to produce other cash crops and the knowledge and the involvement required in setting up and maintaining the CTF system.

Chapter 8 is looking into a pre-commercial and autonomous systems that have the potential to be implemented in the near future, namely robotic crop cultivation systems. The aim of this chapter is to describe the potential economic benefits from the application of agricultural robots under specific conditions and constraints. Examples given are the potential gross margin for early seeding and re-seeding in sugar beet. With some predefined assumptions with regard to speed, capacity and seed mapping, it was found that both early seeding with a small robot and re-seeding with a robot on a small part of the field could be financially viable solutions in sugar beet.

All the above technical systems require that farmers and advisors are able deploy the technologies in a farm management context. Chapter 9 provides a general description of farm management information systems (FMIS). This chapter presents the state-of-the-art depicting the new functionalities included in evolved FMISs and how they can connect the farm to the external context and stakeholders. Subsequently, the authors delve into the functionality of FMIS to understand how precision agriculture can improve the allocation of costs to the final product by managing PA in a better way.

Although many PA technologies are targeted at large scale farming with an aim to increase farm profitability, PA is also a measure to save on inputs and to maintain a sustainable production. Chapter 10 focuses on how precision farming as a sustainable technology can help to increase yields and reduce environmental impact in crop farming; a case study from Estonia is presented in which PF is one of the technologies among others. The application of different technologies used in crop production was analysed and many previously presented factors from the literature about sustainable intensification were described.

Recently, adoption of PA technologies has been addressed in several national as well as multinational studies. Chapter 11 provides an overview of different adoption studies that have been conducted on precision agriculture. It shows that PA has been shown to affect the performance of farms positively, even though its benefits vary according to the size of farms as well as their location. In light of the promising avenue that precision agriculture opens up, it is essential to understand which factors may facilitate its diffusion and through which processes. This chapter focuses on the models proposed to explain technology adoption. Morever, in this chapter some reflections are presented on how to expand knowledge of precision agriculture along this line of reasoning aimed at integrating personal and social characteristics. The importance of social network patterns and of social support in entrepreneurial initiatives that sustain precision agriculture adoption is also highlighted in this chapter.

Chapter 12 examines the perspectives of precision farming in a broader policy context. Agriculture is faced with contrasting opposing requirements from the broader society. On the one hand, agriculture needs to expand production in order to be able to feed a growing global population. Furthermore, the developing bio-economy requires agriculture to produce a range of non-food objectives such as bio-fuel, textile fibres, etc. On the other hand, concerns over the environment, climate, biodiversity and other public goods place restrictions on conventional agricultural production. Precision agriculture can be part of the response to these often conflicting issues by employing technologies that in a precise and targeted approach reduces resource use and increases yield. Furthermore, the growing demand for higher value food products in terms of health and quality properties that require traceability and information about production processes and resource use corresponds with the possibilities offered by precision agriculture technology. The general movement towards higher integration in food supply chains is a natural extension of the requirements for traceability and product information, which are integral parts of precision agriculture.

1.6 Summary and Main Findings

Precision farming has until now been technology driven in which several novel GPS based farm-technologies have been introduced in the farming society to make decisions about the site-specific application of nutrients and pesticides. The development is both driven by private companies and to some extent public interests. Most of the systems are designed to gather information about site-specific crop status and soil conditions and to make better decisions to increase yield or save inputs.

In the last decades, UAV's, auto-steering, controlled traffic farming and most recently autonomous systems in farming have emerged.

Findings from this book show that several of these new technologies and systems appear to be financially viable to implement among farmers. In particular, auto-steering has turned out to be profitable and has been adopted by many farmers. Moreover, UAVs and free of charge satellite images are expected to provide faster and more reliable images of the field conditions at lower costs. In addition, other specific farm management systems like site-specific irrigation appear to be viable as indicated from studies in Australia. Despite all these developments there is still a lack of adoption among farmers, especially small farms, and the economic benefits from variable rate technology still appear to be inconsequential.

One of the main focus areas is variable-rate nitrogen application, which has so far not provided significant returns to investments which is explained by flat response rate functions. To improve the adoption of variable nitrogen application, companies and researchers should continue the development of scientifically sound decision-support systems based on real-time information, soil monitoring, weather forecasts and the field history. Farmers need clear indications that show yield/profit improvements especially from variable-rate fertilizer technologies. To improve the development of PF the following aspects are important:

- More focus should be put on integrated farm management information systems and integrated decision support that include historic yields, soil texture, weather forecast and real time canopy sensing as well as financial variables.
- Need for more efficient image based autonomous systems to detect pests (weed, insects and fungi) in the field in order to provide reliable and cost efficient solutions to farmers.
- Site-specific fertilizer applications should integrate and manage manure, slurry and mineral fertilizers in an integrated fertilizer plan as well as real time information about crop's status.
- Manufacturers should focus on the development of compatible, reliable and low cost hardware that is easy to implement in existing farming systems.
- Better coordination of experiences and establishment of a common database about PF data among farmers.

Although the economic returns from different precision farming technologies have provided mixed results there are still promising perspectives in the light of the latest developments in UAV's, faster computers, better crop models and the introduction of small low cost cameras and sensing systems. All this combined with fast data processing with timely and accurate decision support is likely to be improved further in years to come. In addition, PF will be an important technology as a means to fulfil the current environmental trends that is a part of the regulation of farms across Europe as well as other countries.

Finally, the growing demand for higher value food products in terms of health and quality properties requiring traceability and information about production processes and resource use corresponds with the possibilities offered by precision agriculture technology. The general movement towards higher integration in food supply chains is a natural extension of the requirements for traceability and product information, which are integral parts of precision agriculture.

References

Abidine AZ, Heidman BC, Upadhyaya SK, Hills DJ (2002) Application of RTK GPS based auto-guidance system in agricultural production. Paper No. O21152. ASAE, St. Joseph

Audsley E (1993) Operational research analysis of patch spraying. Crop Prot 12:111–119

Basso B, Bertocco M, Sartori L, Martin EC (2007) Analyzing the effects of climate variability on spatial patterns of yield in a maize-wheat-soybean rotation. Eur J Agron 26:82–91

Basso B, Cammarano D, Troccoli A, Chen D, Ritchie JT (2010) Long-term wheat response to nitrogen in a rainfed Mediterranean environment: field data and simulation analysis. Eur J Agron 33(2010):132–138

Basso B, Ritchie JT, Cammarano D, Sartori L (2011) A strategic and tactical management approach to select optimal N fertilizer rates for wheat in a spatially variable field. Eur J Agron 35:215–222

Basso B, Sartori L, Cammarano D, Grace PR, Fountas S, Sorensen C (2012) Environmental and economic evaluation of N fertilizer rates 1 in a maize crop in Italy: a spatial and temporal analysis. Biosyst Eng 113:103–111

Bongiovanni R, Lowenberg-Deboer J (2000) Economics of variable rate lime in Indiana. Precis Agric 2(1):55–70

Christensen S, Heisel T, Secher BJ (1997) Spatial variation of pesticide doses adjusted to varying canopy density in cereals, Precision agriculture. BIOS Scientific Publishers Ltd, Oxford, pp 211–218

Christensen S, Rasmussen J, Pedersen SM, Dorado J, Fernandez-Quintanilla C (2014) Second International Conference on Robotics and associated High-technologies and Equipment for Agriculture and Forestry. New trends in mobile robotics, perception and actuation for agriculture and forestry. RHEA Project, Madrid, pp 541–549

Daberkow SG (1997) Adoption rates for recommended crop management practices: implications for precision farming. In: Stafford JV (ed) Precision Agriculture 1997, proceedings of the 1st European conference. BIOS Scientific Publishers, Warwick, pp 941–948

Dammer K-H, Adamek R (2012) Sensor-based insecticide spraying to control cereal aphids and preserve lady beetles. Agron J 104(6):1694–1701

Davies S (1979) Technical choise, innovation and economic growth. Cambridge University Press, Cambridge

European Union (2014) Joint Recesrch Cneter precision agriculture: an opportunity for EU farmers – potential support with the CAP 2014–2020. Pablo J Zarco-Tejada, Neil Hubbard and Philippe Loudjani

Franco C, Pedersen SM, Papaharalampos H, Ørum JE (2017) The value of precision for image-based decision support in weed management. Precis Agric 18(3):366–382

Gerhards R, Sökefeld M, Timmermann C, Reichart S, Kübauch W, Williams MM II (1999) Results of a four-year study on site-specific herbicide application. In: Stafford JV (ed) Precision agriculture '99, proceedings of the 2nd European conference on precision agriculture, pp 689–697

Heeje HJ (2013a) Heege JL (ed.) Precision in crop farming: site specific concepts and sensing. Precision in Guidance of Farm Machinery, pp 35–50

Heeje HJ (2013b) Heege JL (ed.) Precision in crop farming: site-specific sowing. Dordrecht/New York, Springer, pp 171–192

Jensen HG, Jacobsen LB, Pedersen SM, Tavella E (2012) Socioeconomic impact of widespread adoption of precision farming and controlled traffic systems in Denmark has now been published in the following paginated issue of. Precis Agric 13(6):661–677

Lawson LG, Pedersen SM, Sørensen CAG, Pesonen L, Fountas S, Werner A (2011) Four nation survey of farm information management and advanced farming systems: a descriptive analysis of survey responses. Comput Electron Agric 77(1):7–20

Leiva FR, Morris J, Blackmore SB (1997) Precision farming techniques for sustainable agriculture. In: Stafford JV (ed) Precision agriculture 1997, proceedings of the 1st European conference. BIOS Scientific Publishers, Warwick, pp 957–966

Miller N, Griffin TW, Bergtold J, Sharda A, Ciampitti I (2017) Adoption of precision agriculture technology bundles on kansas farms. Paper prepared for presentation at the Southern Agricultural Economics Association (SAEA) Annual Meeting, Mobile, Alabama, February 4–7, 2017

Mondal P, Basu M (2009) Adoption of precision agriculture technologies in India and in some developing countries: scope, present status and strategies. Prog Nat Sci 19(6):659–666

OECD (2016) Farm management practices to foster green growth. OECD Publishing, Paris

Oliver M, Bishop T, Marchant B (2013) Precision agriculture for sustainability and environmental protection. Routledge, Abingdon

Pedersen SM (2003) Precision farming – technology assessment of site-specific input application in cereals. Department of Manufacturing Engineering and Management, Technical University of Denmark. Ph.D dissertation, 343 p

Pedersen SM, Pedersen JL (2002) Economic and Environmental impact of Site-Specific N-application – based on different weather conditions and arable crops. Paper presentation at the 6th International Conference on Precision Agriculture, Jul 14–17 2002, Minneapolis

Pedersen SM, Ferguson RB, Lark M (2001) A comparison of producer adoption of precision agricultural practices in Denmark, the United Kingdom and the United States, SJFI – Working Paper No 2. Statens jordbrugs- og fiskeriokonomiske Institut, Copenhagen

Pedersen SM, Fountas S, Blackmore S, Pedersen JL, Pedersen HH (2003) In: Stafford J (ed) Adoption of precision farming in Denmark, precision agriculture. 4th ECPA, Berlin, pp 533–538

Pierce FJ, Nowak P (1999) Aspects of precision agriculture. Adv Agron 67:1–85

Reichardt M, Jürgens C (2008) Adoption and future perspective of precision farming in Germany: results of several surveys among different agricultural target groups. Precis Agric 10(1):73–94

Ritchie JT (1998) Soil water balance and plant water stress. In: Tsuji GY, Hoogenboom G, Thornton PK (eds) Understanding options for agricultural production. Kluwer in cooperation with ICASA, Dordrecht/Boston/London, pp 41–54

Robertson MJ, Lyle G, Bowden JW (2008) Within-field variability of wheat yield and economic implications for spatially variable nutrient management. Field Crop Res 105:211–220. https://doi.org/10.1016/j.fcr.2007.10.005

Schimmelpfennig D (2016) Farm profits and adoption of precision agriculture. A report summary from the Economic Research Service. Washington, DC, USDA

Senthilkumar S, Basso B, Kravchenko AN, Robertson GP (2009) Contemporary evidence of soil carbon loss in the U.S. corn belt. Soil Sci Soc Amer Jour 73(6):2078–2086

Srinivasan A (ed) (2006) Handbook of precision farming: principles and applications. The Haworth Press, Inc, Binghamton, p 684

Swinton SM (2005) Economics of site-specific weed management. Weed Sci 53(2):259–263

Timmermann C, Gerhards R, Kühbauch W (2003) The economic impact of site-specific weed control. Precis Agric 4(3):249–260

van Evert F, van der Voet P, Van Valkengoed E, Kooistra L, Kempenaar C (2012) Satellite-based herbicide rate recommendation for potato haulm killing. Eur J Agron 43:49–57

Wang D, Prato T, Qiu Z et al (2003) Economic and environmental evaluation of variable rate nitrogen and lime application for claypan soil fields. Precis Agric 4:35–52

Chapter 2
Smart Farming Technologies – Description, Taxonomy and Economic Impact

Athanasios T. Balafoutis, Bert Beck, Spyros Fountas, Zisis Tsiropoulos, Jürgen Vangeyte, Tamme van der Wal, I. Soto-Embodas, Manuel Gómez-Barbero, and Søren Marcus Pedersen

Abstract Precision Agriculture is a cyclic optimization process where data have to be collected from the field, analysed and evaluated and finally used for decision making for site-specific management of the field. Smart farming technologies (SFT) cover all these aspects of precision agriculture and can be categorized in data acquisition, data analysis and evaluation and precision application technologies. Data acquisition technologies include GNSS technologies, mapping technologies, data acquisition of environmental properties and machines and their properties. Data analysis and evaluation technologies comprise the delineation of management zones, decision support systems and farm management information systems. Finally, preci-

A.T. Balafoutis (✉)
Agricultural University of Athens, Athens, Greece

Institute of Bio-economy & Agro-technology Centre of Research & Technology Hellas, Volos, Greece
e-mail: abalafoutis@aua.gr

B. Beck • J. Vangeyte
Institute for Agricultural, Fisheries and Food Research (ILVO), Merelbeke, Belgium
e-mail: bert.beck@ilvo.vlaanderen.be; jurgen.vangeyte@ilvo.vlaanderen.be

S. Fountas • Z. Tsiropoulos
Agricultural University of Athens, Athens, Greece
e-mail: sfountas@aua.gr; tsiropoulos@teilar.gr

T. van der Wal
Wageningen Environmental Research (Alterra), Wageningen, The Netherlands
e-mail: Tamme.vanderWal@aerovision.nl

I. Soto-Embodas • M. Gómez-Barbero
European Commission – Joint Research Centre (JRC), Seville, Spain
e-mail: iria.soto-embodas@ec.europa.eu; Manuel.GOMEZ-BARBERO@ec.europa.eu

S.M. Pedersen
Department of Food and Resource Economics, Faculty of Science, University of Copenhagen, Frederiksberg C, Denmark
e-mail: marcus@ifro.ku.dk

S.M. Pedersen, K.M. Lind (eds.), *Precision Agriculture: Technology and Economic Perspectives*, Progress in Precision Agriculture,
https://doi.org/10.1007/978-3-319-68715-5_2

sion application technologies embrace variable-rate application technologies, precision irrigation and weeding and machine guidance. In this chapter, the reader can find a technical description of the technologies included in each category accompanied by a taxonomy of all SFT in terms of farming system type, cropping system, availability, level of investment and farmers' motives to adopt them. Finally, the economic impact that each SFT has compared to conventional agricultural practices is given.

Keywords Precision agriculture • Smart farming technologies • Data acquisition • Variable application • Economic impact

2.1 Precision Agriculture as a Cyclic Optimization Process

Precision agriculture (PA) can be defined as the management of spatial and temporal variation in the fields with regard to soil, atmosphere and plants using information and communications technologies (ICT). Several definitions can be found in the literature (Bramley 2001; Pedersen 2003; Fountas et al. 2005; Zarco-Tejada et al. 2014). Precision agriculture is a management system for farms that aims to improve productivity and resources use either through increased yields or reduced inputs and adverse environmental effects. It can assist crop producers because it enables precise and optimized use of inputs leading to reduced costs and environmental impact, and because the concept provides a record (traceability) of farm activities that consumers and central administration increasingly require (Stafford 2000).

Precision agriculture is not a new idea. A few decades ago the farms were small and the farmer would walk all over the fields several times every year. It was possible to observe all within-field variation and take appropriate management decisions for each part. Adding more seeds in parts where emergence was poor or more fertilizer where growth was weak was the dominant practice. However, this knowledge depended on the farmer's memory, and in most cases the final decisions were influenced more by results from recent years that were kept in memory.But these yields were more influenced by weather or other factors that might not occur during the following years. The connection of the farmer with the fields and the knowledge of their specific characteristics were reduced with the mechanisation and the increase in farm size. The average rule was used to manage the fields. When the first yield monitors were developed and yield maps were created, it was proved that yield and soil properties varied considerably within a field. This fact marked the development of site-specific techniques, which are a core discipline of PA. Site-specific farming can be used for any field or crop for applying treatments to areas within a field that require different management from the field average allowing fine-tuning of crop management systems.

Precision agriculture is a cyclic system of data collection used for crop management and evaluation of the decisions, with the cycle continuing for the subsequent years. Each year data are stored in a database (library) and are used as historical data for future decisions. The system can be divided into data collection, data analysis, managerial decisions and variable-rate applications (VRA) of inputs, evaluation of the managerial decisions and a new cycle starts.

To apply the PA cyclic system, there is a need for a series of technologies called smart farming technologies (SFTs) that refer to marketable, affordable, reliable and time-saving technologies drawing from research in precision farming, farm management information systems (FMIS) and agricultural automation and robotics. Their benefits are related to more efficient application of inputs (seeds, fertilizers, chemicals, water, fuel and labour), increased work speed, comfort and enhanced flexibility on the farm.

The cyclic production process can be achieved by data acquisition, data analysis and evaluation (decision making) and precise application of operations (field implementation). Therefore, the SFTs presented in this chapter are classified based on this structure.

2.2 Smart Farming Technology Types

Smart farming technologies (SFTs) are divided into three main categories that, as stated above, cover the cyclic system of PA:

- **Data acquisition technologies**: this category contains all surveying, mapping, navigation and sensing technologies.
- **Data analysis and evaluation technologies**: these technologies range from simple computer-based decision models to complex farm management and information systems including many different variables.
- **Precision application technologies**: this category contains all application technologies, focusing on variable-rate application and guidance technologies.

There is a series of technologies that can be classified in each category of SFTs, as shown in Table 2.1.

Each technology referred in the table above will be analysed in this chapter.

2.2.1 Data Acquisition Technologies

The SFTs for recording and mapping field and crop characteristics are divided into the categories below:

- Global navigation satellite systems technologies (in fact these technologies record the actual position which can be used for different purposes such as guidance, mapping etc.)
- Mapping technologies
- Data acquisition of environmental properties (Camera based imaging, NDVI measurements, soil moisture sensors)
- Machines and their propertiesGlobal navigation satellite systems (GNSS) technologies

Table 2.1 Smart farming technologies list

PA technologies	Main categories	System
Data acquisition technologies	GNSS technologies	Global navigation satellite systems (GNSS)
		Differential GNSS
		Real time kinematic (RTK) and Network RTK (NRTK)
		Wide area RTK (WARTK)
		Un-differenced GNSS
		Precise point positioning (PPP)
		Fast PPP (FPPP)
	Mapping technologies	Elevation maps
		Soil mapping
		Yield mapping
		Yield monitor display
	Data acquisition of environmental properties echnologies (Camera based imaging)	RGB cameras
		LIDAR sensors
		ToF (IR) cameras
		Light curtains
		Multi/hyper-spectral cameras
		Thermal cameras
	Data acquisition of environmental properties technologies (NDVI Measurement)	Spectral sensors
		Fluorescence sensors
	Data acquisition of environmental properties technologies (Soil moisture sensors)	Frequency domain reflectometry (FDR)
		Time domain reflectometry (TDR)
		Amplitude domain reflectometry (Impedance)
		Phase transmission
		Time domain transmission
		Tensiometers
		Gipsum blocks
		Granular matrix sensors
		Heat dissipation sensors
	Machines and their properties	Travel speed sensor
		Tractor sensing systems using ISOBUS
		Unmanned aerial vehicles (UAVs)
		Unmanned ground vehicles (UGVs)
		Farm management information system
		Software for whole farm management, forecasting and crop monitoring

(continued)

Table 2.1 (continued)

PA technologies	Main categories	System
Data analysis & evaluation technologies		Management zone delineation
		Decision support system
Precision application technologies	Guidance technology	Auto-guidance systems
		Control traffic farming
	Variable rate application	Variable-rate fertilizer application
		Variable-rate lime spplication
		Variable-rate manure application
		Variable-rate pesticide application (Map-based system)
		Variable-aate pesticide application (Real-time sensor based system)
		Boom height control
		Variable-rate planting/seeding
		Precision physical weeding
		Precision irrigation and irrigation scheduling

2.2.1.1 Global Navigation Satellite Systems (GNSS)

Global navigation satellite systems (GNSS) is the standard generic term for satellite navigation systems that provide autonomous geo-spatial positioning with global coverage. Any GNSS is used to pinpoint the geographic location of a user's receiver anywhere in the world. Currently, there are two operational GNSS systems (GPS and GLONASS) and two systems in development (Galileo and BeiDou) that are expanding their coverage from regional coverage to global; both are expected to be fully functional in 2020. A brief summary of these different GNSS systems is presented in Table 2.2.

All GNSSs at the moment use over 70 satellites, but when all four systems (GPS, GLONASS, Galileo and BeiDou) are fully deployed it will reach 120 satellites (Li et al. 2015). The GPS has six orbital planes with a minimum of three satellites per plane (28–31 satellites that are constantly working). GLONASS has three orbital planes with eight satellites per plane (24 satellites in total). Galileo will have three orbital planes and a total of 30 satellites.

The precision of GNSS varies. For example, GPS signals originally used an intentional degradation (known as Selective Availability, SA) to prevent potential military adversaries from using the positioning data (military operated system). Therefore, GPS accuracy was limited to a 100-m range for civilian users, although military equipment enabled accuracy to within a metre. In May 2000, SA was discontinued and since then all GPS receivers are potentially accurate to within 5 m.

Table 2.2 Main GNSS systems

System	GPS	GLONASS	Galileo	BeiDou
Owner	United States	Russian Federation	European Union	People's Republic of China
Site	www.gps.gov	www.glonass-iac.ru	www.gsa.europa.eu	en.beidou.gov.cn
Type	Military	Military	Civilian Commercial	Military Commercial
Coding	CDMA	FDMA	CDMA	CDMA
Orbital altitude	20,180 km	19,130 km	23,222 km	21,150 km
Precision	5 m	3–10 m	1 m	10 m
			0.01 m encrypted	0.1 m encrypted
Period	11.97 hours	11.26 hours	14.08 hours	12.63 hours
N° satellites	**31** (at least 24 by design)	**28** (at least 24 by design),including:24 operational	**4** in-orbit validation satellites +**8** full operation capable satellites in orbit	**5** geostationary orbit (GEO) satellites
		2 under check by the satellite prime contractor	**22** operational satellites budgeted	**30** medium Earth orbit (MEO) satellites
		2 in flight tests phase		
Status	Operational **1995**	Operational **2015**	In development (expected full deployment by **2020**)	Operational (*regionally* **2012**)
				In development (global use by **2020**)

GLONASS showed an accuracy of 35 m in 2006, but after its modernization it reached less than 3 m, which is very similar to GPS. As of 23 January 2012, GLONASS's horizontal precision is in the order of 4–7 m, whereas the vertical error is in the order of 10–15 m. However, analysing the accuracy obtained with GPS at the same stations it has been shown that GLONASS is slightly less accurate than GPS. In the same way, the mean number of GLONASS satellites in view is fewer than GPS.[1]

When available, Galileo will provide position accuracy to within one metre for public use and 1 cm in the encrypted state. The first Galileo test satellite, the GIOVE-A, was launched on December 28 2005, while the first satellite to be part of the operational system was launched on October 21 2011. As of December 2015, the system will have 12 of 30 satellites in orbit and started offering early operational capability (EOC) from 2016 and will go to initial operational capability (IOC) in 2017–2018 and reach full operational capability (FOC) in 2019 (Galileo's contribution to the MEOSAR system, 2015). The complete 30-satellite Galileo system (24 operational and 6 active spares) is expected by 2020.[2]

2.2.1.2 GNSS Precise Positioning Techniques: Differential GNSS

Differential GNSS (DGNSS) is a GNSS augmentation system based on improving the accuracy of the user's receiver (or rover receiver) by means of differential information or corrections provided by a nearby reference GNSS station or a network of these stations. The application of this concept allows common sources of error between satellites and receivers to be cancelled or mitigated, because of dual-frequency carrier-phase measurements and the applyication of double-difference processing.

Classical DGNSS

In the DGNSS approach, we take advantage of knowing an accurate surveyed position of the reference station. In this way, it is possible to derive the deviations between the estimated position and the actual one, and thus compute corrections to the GNSS pseudoranges of each satellite. Such corrections are then useful to improve the user's receiver positioning.

Real Time Kinematic (RTK) and Network RTK (NRTK)

The RTK positioning system was introduced by Remondi (1985). It consists of a user receiver that benefits from a base receiver, with well-known coordinates, and a communication link between both to receive and use the common satellites-in-view measurements to perform the corresponding differences in order to achieve centimetre level positioning accuracy with a short convergence time (Landau et al. 2007).

---

[1] http://www.navipedia.net/index.php/GLONASS_Performances

[2] http://www.gsa.europa.eu/galileo/programme

Wide Area RTK (WARTK)
The WARTK technique, introduced 17 years ago and developed by IonSAT members under several ESA-funded projects, can be considered an extension of RTK/NRTK techniques to enable subdecimetre positioning accuracy with roving receivers hundreds of kilometres away from the reference receiver. To enable this, it is necessary to take as the basic observation the double differences of carrier phases and use additional specific corrections (namely very precise ionospheric Slant TEC estimations) computed at a central processing facility (CPF) from a permanent network of GNSS receivers (Hernández-Pajares et al. 2000).

2.2.1.3 GNSS Precise Positioning Techniques: Undifferenced GNSS

Undifferenced GNSS is a GNSS augmentation system to provide high precision positioning to a user's receiver in absolute mode (i.e. without the need to receive the direct measurements taken from any reference receiver or network of receivers nearby). Instead of that an estimate of specific corrections for satellite orbits and clocks, and ionospheric corrections, among others, is broadcast. As was the case with DGNSS techniques, it is also based on dual-frequency carrier-phase measurements. The application of this concept allows common sources of error between satellites and receivers to be cancelled or mitigated, by using dual-frequency carrier-phase measurements and applying double-difference processing.

Precise Point Positioning (PPP)
Real time PPP (Héroux and Kouba 1995; Zumberge et al. 1997) can be provided in a reliable way by means of a world-wide sparse reference network in order to compute precise reference satellite orbits and clock features in real-time at a CPF. Its architecture allows the applicability of PPP to any user located in a global reference frame without being referred to any local base station or network of stations. In addition, the technique can diminish considerably the impact of failures of certain reference stations by considering a significant number of permanent receivers in order to derive the precise orbit and clock data.

Fast PPP (FPPP)
Fast PPP technique is an evolved version of the classic PPP to achieve decimetre level positioning and also faster convergence time (for double-frequency user receivers) in undifferenced mode. This means that the user navigates without the need for a reference receiver (single receiver navigation).

GNSS Antenna (Receives Satellite Signal)
When a GNSS antenna is applied, then the whole system converts to a mapping system. Therefore, the data received by all sensors are combined with position for every time interval set (average is 1 Hz, but some accurate GNSS receivers used in kinematic applications can reach a frequency of 2–50 Hz and even higher up to 100 Hz) (Yigit 2016).

2.2.2 *Mapping Technologies*

2.2.2.1 Elevation Maps

Elevation is a critical layer in PA because it is very useful to help farmers understand yield response. It influences soil formation, water movement and cropping aspects (Whelan and Taylor 2013). It can determine waterlogged areas, erosion risk, drainage restrictions, and often is related to soil type.[3] Using data from GNSS receivers, it is possible to produce a digital elevation model (DEM) of a field or a farm that can be used to classify terrain characteristics such as slope, aspect, curvature, solar radiation interception, landscape water flow directions and topographic wetness indices. Elevation maps can help to identify how topography can affect agronomic results in a field and of course to level the field (Whelan and Taylor 2013). Using this information, it is possible to produce (i) contours and topography maps, (ii) 3-D modelling of ponding risk, runoff and velocity maps, (iii) farm layout designs, (iv) contour bank design, drainage plans and on-ground implementation and (v) cut and fill land levelling designs.

2.2.2.2 Soil Mapping

Soil sampling is vital to collect information about soil texture (sand, silt, clay contnets), availability of nutrients for crops to grow (N, P, K, Ca, Mg, pH, lime) and other soil chemical properties (organic matter, salinity, nitrate, sulphate, heavy metals) (Foth and Ellis 1988). In addition, it can be used to identify soil compaction, moisture content and other mechanical and physical soil properties. Soil sampling can be executed using the random, adaptive or grid technique. In random sampling, soil cores are obtained from random locations within the field. In adaptive sampling, selected sample locations depend on prior information. Grid sampling involves systematically collecting samples from predetermined points in the field. None of the existing soil sampling practices has been recognized as the most effective (Wollenhaupt et al. 1997).

Another method to map a field's soil properties is the use of on-the-go sensors that have the potential to provide benefits from the increased density of measurements at a relatively low cost. These sensors can be either combined with a GNSS receiver and produce maps of soil properties or they can be used as real-time sensors where the output of the sensor is used immediately for variable-rate application of fertilizers, lime and manure (Fig. 2.1).

[3] http://www.precisionagriculture.com.au/topography-and-drainage.php

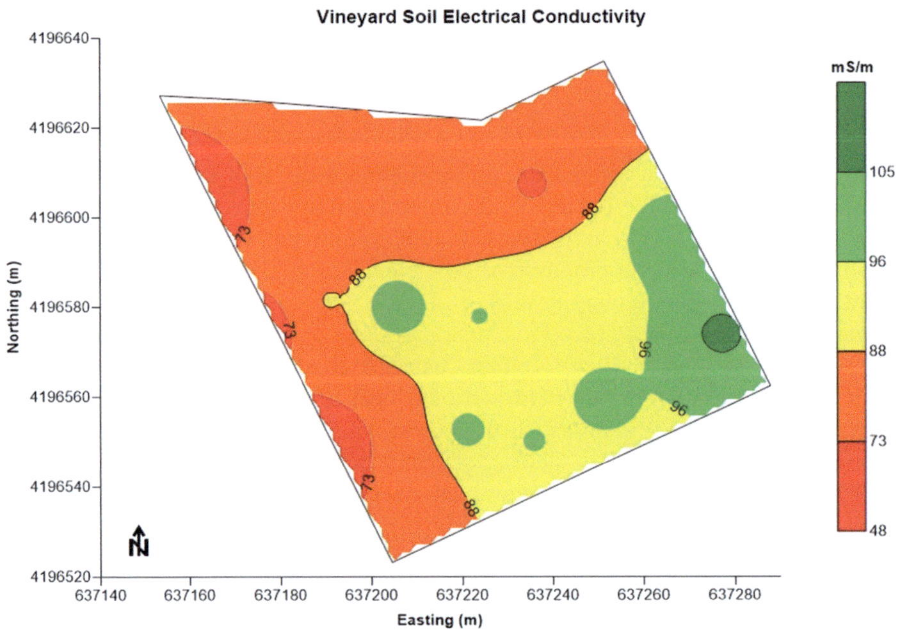

Fig. 2.1 Soil electrical conductivity (ECa) map (Source: Fountas et al. 2014)

There are different kinds of on-the-go soil sensors which can indicate different agronomic soil properties (Adamchuk et al. 2004; Adamchuk and Viscarra Rossel 2014):

- Electrical and electromagnetic sensors measure electrical resistivity or conductivity, capacitance or inductance affected by the composition of the soil. The most common instruments used in research and practice are the EM38 (Geonics, Canada) and the VERIS (VERISTech, USA).
- Mole gamma radiometer (The Soil Company, Groningen, The Netherlands) for predicting clay percent and CEC (cation exchange capacity) of soil.
- Optical and radiometric sensors use electromagnetic waves to detect the level of energy absorbed or reflected by soil particles.
- Mechanical sensors measure forces resulting from a tool engaged with the soil.
- Acoustic sensors quantify the sound produced by a tool interacting with the soil (ex. horizontal penetrometers).
- Pneumatic sensors assess the ability to inject air into the soil.
- Electrochemical sensors use ion-selective membranes that produce a voltage output in response to the activity of selected ions (H^+, K^+, NO_3^-, Na^+, etc.)

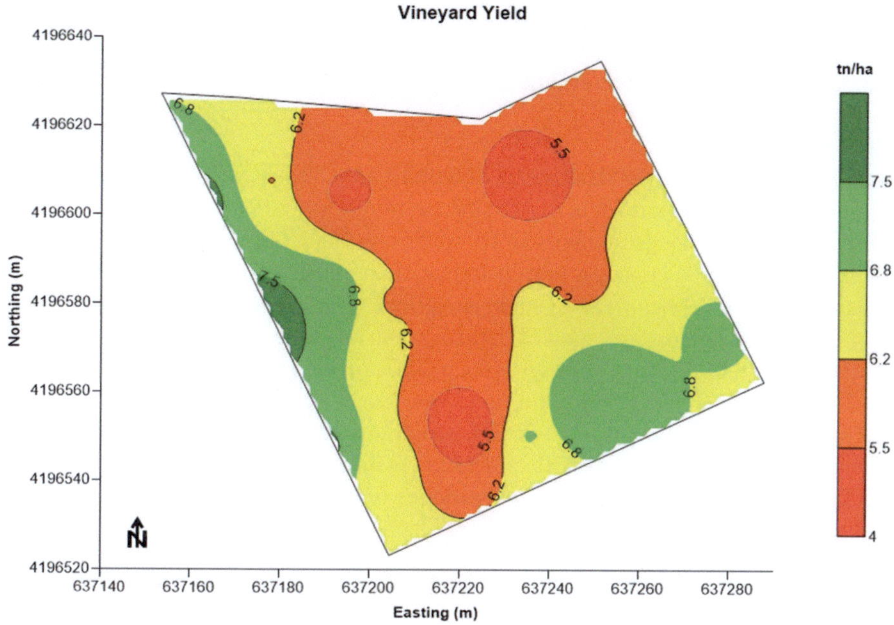

Fig. 2.2 Yield maps (Source: Fountas et al. 2014)

2.2.2.3 Yield Mapping

Yield mapping or yield monitoring is a technique in agriculture of using GNSS data to analyse variables such as crop yield and moisture content in a given field (Fig. 2.2). The components of a grain yield mapping system include a grain flow sensor that measures grain volume, a grain moisture sensor that quantifies moisture variation, a grain elevator speed sensor that measures grain speed to calculate grain mass, a GNSS antenna that geo-references grain measurements, a header position sensor that initiates grain measurement when the header is lowered and a travel speed sensor that provides the distance that the harvester has covered during a certain logging interval.

There are many types of grain sensors that are commercially available, such as a paddle wheel volume flow sensor, momentum plate sensor, gamma ray sensor, strain gauge based impact sensors, infrared sensor. Other yield sensors are also found in literature that are not commercially available, such as pivoted auger, piezo-film strips, capacitive sensor, ultrasonic sensor, elevator based flow sensor, X-ray techniques.

Yield Monitor Display with a GNSS Receiver (Georeference and Record Data)

It is a tablet-type screen in the harvester cabin combined with a processor, data inputs and storage capabilities that allows the operator to import filed information, calibration functions, visual sampling display of the yield and moisture (Whelan and Taylor 2013).

2.2.3 *Recording of Environmental Parameters*

2.2.3.1 Camera Based Imaging

RGB Cameras

Red, Green and Blue (RGB) cameras combine the colours red, green and blue to depict the range of colours that exist in the environment and in the agricultural fields. There is a series of measurements and correlations that RGB images can be used for. Vollmann et al. (2011) used a digital camera Sony DSC F707 (Sony Corp., Tokyo, Japan) to study the phenotype of soybean varieties. Through the use of digital image analysis, a significant correlation of the red, green and blue of digital images with the protein content of soybean plants was found. Thorp and Dierig (2011) used the camera EOS Digital Rebel XT (Canon Inc., New York, USA) for counting the flowers and the whole course of flowering in *Lesquerella fendleri* (Fendler's bladderpod). Wang and Li (2014) used an RGB camera of a Kinect sensor (Microsoft Inc., Seattle, USA) to measure the diameter of two varieties of onions.

LiDAR Sensors

LiDAR sensors (Light Detection and Ranging) are instruments that measure the distance from the target by laser. This technology has been used to study the phenotypic variation by creating three-dimensional models of plants. The principle of LiDAR devices is that they send rapid pulses of laser light to a surface and a sensor on the instrument measures the amount of time it takes for each pulse to bounce back. As the velocity of light is known, the LiDAR devices can calculate the distance between them and the target with high accuracy. When LiDAR is used as a ground sensor, it is required to have a GNSS receiver for the location of the device. As for airborne LiDAR devices, things become more complex when it is required that the moving height, location and orientation of the device are known to determine the position of the laser pulse at the time of sending and the time of return. Generally, there are two types of LiDAR detection methods. "Direct energy detection", also known as incoherent, which is principally an amplitude measurement and "Coherent detection" that are best for Doppler[4] or phase sensitive measurements and generally use optical heterodyne detection. This allows them to operate at much lower power, but has the expense of more complex transceiver requirements. They consist of a laser that produces the beam, a scanner and optics that scan the beam, a photodetector and receiver to receive the beam after its reflection and GNSS receiver with horizontal and vertical accuracies of <3 cm and <15 cm, respectively, to obtain the location of the sensor (Reutebuch et al. 2003).

LiDAR can be used in agricultural applications, such as the creation of topographical maps, slope and sun exposure of the farm. Another application of LiDAR is crop mapping in orchards and vineyards. Foliage growth can be measured to determine if pruning or any other agricultural practice is required, detect variations in fruit

[4] The Doppler Effect is the difference between the observed frequency and the emitted frequency of a wave for an observer moving relative to the source of the waves.

production or perform automated tree counts. Also tree area index (TAI) and leaf area index (LAI) can be estimated with ground LiDAR sensors (Arnó et al. 2013, 2015). For vehicle-based determination of crop biomass, commercially available laser scanners have been analysed and tested to measure aboveground biomass in oilseed rape, winter rye, winter wheat, oats and grassland (Ehlert et al. 2010). Laser scanners are also used for crop height detection (Hoffmeister et al. 2016). Paulus et al. (2014) used the sensor ScanWorks v5 Perceptron (Hexagon Metrology Inc., Plymouth, USA) to create three-dimensional models of barley plants and organs. In this way they measured leaf area, stem height, plant height and thickness of the plant. Hosoi and Omasa (2012) used the sensor TDS-130 L (Pulstec Industrial Co., Ltd., Japan) to calculate the density of winter wheat plants at different growth stages with good results ($r^2 = 0.95$). From the same experiment they calculated the biomass of plant organs (with $r^2 = 0.94$–0.96). Hosoi and Omasa (2012) conducted the same experiment in rice plants. The biomass of the organs of rice was calculated by a LiDAR sensor showing a strong correlation with the direct measurement ($r^2 = 0.94$–0.99). Rosell et al. (2009) used the sensor LIDAR SICK LMS 200 (SICK AG, Waldkirch, Germany) and they were able to reproduce three-dimensional models of fruit trees such as pear, apple, orange and tangerine in real orchards. The correlation showed that three-dimensional models were strongly correlated with the real ones (up to $r^2 = 0.976$). They say that the three-dimensional models can be used for the calculation of height, volume, thickness, leaf area index of the tree and for other traits.

ToF (IR) Cameras

Time of Flight (ToF) cameras have the ability to produce shaped and incoherent infrared light in the space. Smart sensors at pixels of the camera record the reflected light and calculate the time to return. In this way a three-dimensional model is produced (Verdu et al. 2013). Nakarmi and Tang (2012) used the ToF camera SwissRanger SR4000 (Mesa Imaging AG, Zurich, Switzerland) to measure the distance between the corn plants in a row. Their method showed strong correlation ($r^2 = 0.95$) with the actual distance of maize plants on the row. Wang and Li (2014) calculated the diameter and volume of onions using the Kinect sensor (Microsoft Inc., Seattle, USA) with accuracy of measurements of around 96%. Chene et al. (2012) used the same sensor to measure the curvature, the morphology and the leaf orientation of a rosebush.

Light Curtains

Light curtains are a new system that is used to study the phenotypic traits. The system consists of a couple of bars which are placed in parallel. One bar emits light beams that end up at the other parallel bar. In this way the system records if the light beams are blocked by an object. Fanourakis et al. (2014) used light curtains (INFRASCAN 5000, Sitronic GmbH, Austria) to measure the height and leaf area of corn, tomato, barley and oilseed rape plants. Montes et al. (2011) used the light curtains KONTURflex (Leuze electronic GmbH + Co.KG, Owen, Germany) to measure biomass of 10 hybrids and 10 varieties of corn. They found a strong correlation between the biomass and the results light curtains gave ($r^2 = 0.82$–0.87).

Multi- or Hyper-Spectral Cameras

Multispectral cameras are cameras that can photograph the environment with a limited number of spectra in the visible and infrared spectrum. Thus, the normalized difference vegetation index (NDVI) can be produced by this method, which can be used to calculate biomass, distinguish different plant species, maturation of plant, nutrient status, efficiency of photosynthesis or water content and to detect diseases and insect pests. Hyperspectral cameras, as opposed to multispectral cameras, can produce images at hundreds of positions of the electromagnetic spectrum. As a result, hyperspectral cameras can produce a larger number of vegetation indices than multispectral ones. Liu et al. (2014) used the VideometerLab (Videometer A/S, Horsholm, Denmark) to calculate quality properties such as consistency, concentration of sugars and ripening in strawberries. They found a correlation of $r = 0.94$ for the consistency of strawberry and $r = 0.83$ for the concentration of sugars in relation to the actual values. Zarco-Tejada et al. (2013a) used the hyperspectral camera Micro-Hyperspec VNIR (Headwall Photonics, MA, USA), which was adapted onto a UAV to calculate carotenoids in vineyards with very good correlation ($r^2 = 0.84$). Berni et al. (2009) used multispectral camera MCA-6 (Tetracam Inc., CA, USA) in an unmanned aerial vehicle (UAV) to produce vegetation indices in corn plants and olive trees. Calderon et al. (2013) used a multispectral (MCA-6, Tetracam Inc., California, USA) and a hyperspectral camera (Micro-Hyperspec VNIR, Headwall Photonics, MA, USA), which were mounted on a UAV to locate olives infected by the pathogen *Verticillium* wilt through various vegetation indices.

Thermal Cameras

Thermal cameras have the ability to generate images related to the ambient temperature. This is because they work in the long wavelength infrared (to 14,000 nm) resulting in perceiving the radiation emitted by the target because of its heat. Thermal cameras have been used to study the phenotypic variance for predicting water stress of plants, to detect diseases and pathogens and for the ripening of fruits. Zarco-Tejada et al. (2013b) used a thermal camera (Miricle 307, Thermoteknix Systems Ltd., Cambridge, UK) to study the water stress of vines through the index CWSI and found a strong correlation with the water potential of the leaves ($r = 0.95$). Benavente et al. (2013) used the thermal camera FLIR SC305 (Inframetrics, FLIR Systems Inc., OR, USA) to assess the durability of various genotypes of *Brachypodium distachyon* and *Brachypodium hybridum* in drought tolerance. They found that the genotypes of the species *Brachypodium Hybridum* showed greater resilience to drought, which they related to the better functioning of stomata of this species to water stress in relation to the *Brachypodium Distachyon* species.

2.2.3.2 Normalized Difference Vegetation Index (NDVI)

The most important application for precision agriculture is measuring vegetation indices and more particularly NDVI (Fig. 2.3). The NDVI is a numerical index based on the visible and near-infrared bands of the electromagnetic spectrum that

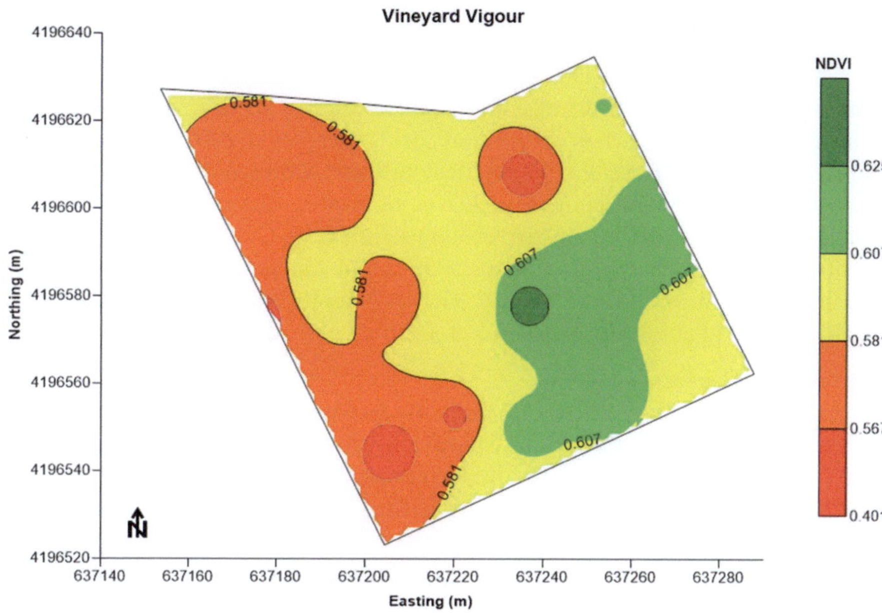

Fig. 2.3 NDVI map (Fountas et al. 2014)

indicates if a target being observed contains live green vegetation or not; it takes values between 0 and 1.

There are many applications of NDVI for either agriculture or environmental solutions. It can be used to estimate crop yields, percentage ground cover, photosynthetic activity of the plant, surface water, leaf area index, the amount of biomass, pasture performance, rangeland carrying capacities, etc. The NDVI was first used in 1973 by Rouse et al. (1973) from the Remote Sensing Centre of Texas A&M University.

There are several commercial products that can be used for NDVI measurements on ground (active) sensors based on the principle of light emission. More particularly, they emit light towards the plant canopy in visible (VIS) and near infra-red (NIR) light that is either reflected, transmitted or absorbed. According to the plant characteristics, the percentage of each of the three behaviours of the light is differentiated (Inman et al. 2005).

On-the-go NDVI ground sensors can be combined with either a GNSS receiver to produce maps of NDVI or they can be used as real-time sensors where the output of the sensor is used immediately for variable-rate fertilizer or spraying applications.

Spectral Sensors

Spectral sensors are instruments that can sense the amount of light reflecting from objects, which they convert to an electrical signal. They measure light in the visible (400–700 nm) and infrared spectrum (700–2500 nm). Spectral sensors are used widely in agriculture because it has been found that these measurements are related to a plant's physiology and development. These sensors may be either passive or active. The passive spectral sensors use sunlight, whereas active sensors have light sources that generate radiation for conducting the measurements. For this reason, the active spectral sensors are less dependent on weather conditions. There are many types of spectral sensors of both categories such as spectrometers, spectroradiometers and canopy sensors (Erdle et al. 2011). Feng et al. (2008) used the spectrometer ASD Field Spec Pro (Analytical Spectral Devices, CO, USA) for the measurement of nitrogen content of leaves of winter wheat varieties. Andrade-Sanchez et al. (2014) used the active canopy sensor Crop Circle ACS-470 (Holland Scientific, NE, USA) to evaluate vigour of 25 cotton varieties. Ramirez et al. (2014) used the spectral instrument SPAD-502 (Konica Minolta, Osaka, Japan) for measuring chlorophyll to assess the resistance of a variety of potato in drought tolerance.

Fluorescence Sensors
Fluorescence induced by ultraviolet radiation has been used as a non-destructive method for estimating plants status. Specifically, the fluorescence of plants caused by UV radiation has been used for the identification of species of plants, for plant growth, for lack of nutrients in plants, for lack of water, for temperature effects on plants and for detecting attacks by pathogens of plants (Cerovic et al. 1999). Thus, various types of sensors have been developed to study fluorescence of plants. Christen et al. (2007) used the fluorometer Handy-PEA (Handy-Plant Efficiency Analyser, Hansatech Instruments, Norfolk, UK) for the detection of *Esca* disease in vines and the results were compared to water stress. Thoren et al. (2010) used the sensor N-detector (Planto GmbH, Germany) to study the fluorescence caused by different common wheat plant fertilizers. Ghozlen et al. (2010) used the optical fluorescence sensor Multiplex (FORCE-A, Orsay, France) to measure the content of anthocyanins in the red grape variety Champagne by a non-destructive method.

2.2.3.3 Soil Moisture Sensors

Information on the spatial and temporal evolution of soil moisture is of great importance for the use of soils and for vegetation, in particular where the water resources are scarce. There are several reliable ways to measure soil moisture. Various *in situ* sensors are available and suitable for precise and reliable measurement of soil moisture (Munoz-Carpena 2017).

Frequency Domain Reflectometry (FDR) Sensors (Capacitance)
When a capacitor uses the soil as a dielectric, its electrical capacitance depends on the soil water content. Such capacitors can be made of metal plates or rods. If this capacitor type is connected with an oscillator to form an electrical circuit, any change in the circuit's operating frequency indicates changes in soil moisture. This is the working principle of capacitance and frequency domain reflectometry (FDR) sensors.

These sensors have probes that consist of two or more electrodes, both inserted into the soil. If an electrical field is applied, the oscillating circuit is completed by the formation of the dielectric of the capacitor by the soil around the electrodes. It is possible to use an access tube that allows installation of multiple sensors at different soil depths. Soil-specific calibration of these sensors is recommended because of their low operating frequency (below 100 MHz), which affects the bulk permittivity of soil minerals and properties such as temperature, salinity, bulk density and clay content may change the measurement.

Therefore, these sensors are accurate after soil-specific calibration, they can read in high salinity levels, they offer better resolution than TDR (see below), can be connected to conventional loggers, are flexible in probe design and are relatively inexpensive. However, their sensing sphere of influence is relatively small, it is extremely critical to have good contact between these sensors and soil, careful installation is necessary to avoid air gaps. They tend to have greater sensitivity to temperature, bulk density, clay content and air gaps than TDR and they require soil-specific calibration.

Time-Domain Reflectometry (TDR) Sensors

These sensors are based on measuring the time it takes for an electromagnetic pulse (wave) to propagate along a transmission line surrounded by the soil. Therefore, TDR sensors produce a series of precisely timed electrical pulses with a wide range of high frequencies, that travel along a transmission line that is built with a coaxial cable and a probe. In contrast to FDR sensors, the high frequency of operation makes measurements less dependent on soil-specific properties such as texture, salinity or temperature. The TDR sensors usually have probes consisting of 2–3 parallel metal rods that are inserted into the soil acting as waveguides, while they have a device for measuring and digitizing the energy (voltage) level of the transmission line at intervals of around 100 picoseconds. As the electromagnetic pulse travels along the transmission line, it faces a discontinuity (i.e. probe-waveguides surrounded by soil) and a part of the pulse is reflected, producing a change in the energy level of the transmission line.

These sensors are accurate, they do not require soil specific-calibration (with minor exceptions), they can easily be expanded by multiplexing, they have minimal soil disturbance and they can provide simultaneous measurements of soil electrical conductivity. However, they are relatively expensive because of the complex electronics, they are not good for highly saline conditions or in strongly conductive heavy clay soils, they need to be calibrated for some soil types (with large amounts of bound water, with large organic matter content, volcanic soil, and so on) and they have a relatively small sensing volume.

Amplitude Domain Reflectometry (Impedance)

The working principle of these sensors is based on the reflection of a part of the energy transmitted (electromagnetic wave travelling along a transmission line) back to the transmitter when the wave reaches a section with different impedance. A voltage standing wave along the transmission line is produced when the reflected wave interacts with the incident wave. These sensors minimize the effect of soil electrical conductivity by choosing a signal frequency so that soil water content can be estimated from the soil or probe impedance.

An electromagnetic wave at a fixed frequency is generated by an oscillator to be applied to a coaxial transmission line that continues in the soil through parallel metal rods that have an electrical shield in the outer part and a central signal rod.

The impedance of this rod arrangement depends on the dielectric constant of the soil between the rods.

These sensors are accurate with soil-specific calibration, enable measurements in very saline conditions, produce minimal soil disturbance, can be connected to conventional loggers, are inexpensive, are not affected by temperature and can estimate soil bulk density. However, they have a small sensing volume, therefore, it is recommended to calibrate them for reliable measurements and their measurements are affected by air gaps, stones or water channeled directly on to the probe rods.

Phase Transmission

The principle of these sensors is based on the phase shift that an electromagnetic wave at a fixed frequency will express in relation to its phase at the origin after travelling a fixed distance. The properties that produce this phase shift are the length of travel along the transmission line, the frequency and the velocity of propagation. Therefore, knowing that velocity of propagation is related to soil moisture content, when a fixed frequency is used and the length of travel is stable, soil water content can be determined by this phase shift. The probe of these sensors consists of two open concentric metal rings to apply phase measuring electronics at the beginning and end of the wave guides.

These sensors are very accurate with soil-specific calibration, they have large soil sensing volume, they can be connected to conventional data loggers and they are inexpensive. However, they cause considerable soil disturbance during installation because of the concentric rings sensor configuration, require soil-specific calibration, are sensitive to salinity levels, have reduced precision because the pulse generated gets distorted during transmission and it needs to be installed permanently in the field.

Time Domain Transmission

These sensors measure the time that an electromagnetic pulse requires to propagate along a transmission line (one-way). They are similar to TDR sensors, but in this case an electrical connection at the beginning and end of the transmission line is needed. The probe consists of bent metal rods to achieve the insertion at the beginning and end of the transmission line in the electronic block.

These sensors are accurate, have large sensing soil volume, can be connected to conventional loggers and are inexpensive. However, they have reduced precision because the pulse generated is distorted during transmission, it disturbs the soil during installation and need to be installed permanently in the field.

Tensiometers

Tensiometers are based on the principle of water equilibrium between the soil solution and the water content of a sealed water-filled tube installed in the soil through a permeable and saturated porous material. This equilibrium results from achieving the same pressure potential for both the water in the tube and the water held in the soil matrix. Hence, the soil water matric potential is equivalent to the vacuum or suction created inside the tube. These sensors consist of a sealed water-filled plastic tube with a ceramic cup at one end and a negative pressure gauge at the other. There are many shapes and sizes of the ceramic cup and the accuracy can vary depending on the gauge or transducer used. Tensiometers for fine soil have a measurement range between 0–80 kPa but for coarse soil there are low-tension versions (0–40 kPa).

Tensiometers provide direct readings up to a 10 cm measurement sphere radius, continuous is reading possible when a pressure transducer is used, avoid electronics and power consumption, are well-suited for high frequency sampling or irrigation schedules, need minimal skill for maintenance, not affected by soil salinity and are inexpensive. However, they have limited soil suction range (<100 kPa), a relatively slow response time, require intimate contact with soil around the ceramic cup for consistent readings and require frequent maintenance (refilling) to keep the tube full of water, especially in hot dry weather.

Gypsum Blocks

These sensors determine soil moisture by measuring the resistance between two electrodes inside the gypsum blocks, which is proportional to water content of the block (low resistance when water content gets smaller). Gypsum blocks are porous, so their water content is related to the moisture of the soil that in which it is being installed. The condition for reliable measurements is optimal contact between sensor and soil. The gypsum blocks are buried permanently in the soil at the desired depth with a life expectancy of 3 to 5 years (depending on the type of soil). The meter is practical and is constructed of sturdy synthetic material. It has a measuring range of 0–100% for 3–100 kPa of water pressure and is applied in places where a typical tensiometer cannot be used (dry soil). The advantages of these sensors are that they are simple and cheap with a large measurement range, they do not need maintenance and they are well suited to area where the soil becomes dry (where there are trees). The disadvantages are that they have low resolution, they react quickly on wetting, but are slow on drying, they are temperature dependent and they are not frost and salt resistant. In addition, the fact that gypsum dissolves over time in different ways for each sensor installed in a field = means that it does not maintain the quality of measurement over time and from site to site.

Granular Matrix Sensors

Water conditions inside granular matrix sensors change with corresponding variation in water conditions in the soil.[5] These changes within the sensor are reflected by differences in electrical resistance between two electrodes imbedded in the sensor. Resistance between the electrodes decreases with increasing soil water. These sensors have a porous ceramic external shell with an internal matrix structure that contains two electrodes. An internal cylindrical gypsum tablet buffers against soil salinity levels that occur in some types of irrigated soil. A synthetic porous membrane is surrounded by a stainless steel casing or sleeve with holes. In this case there is a transmission material that is used to respond to soil wetting and drying cycles.

These sensors are simple and inexpensive; they do not dissolve in the soil over time like gypsum blocks and they do not need maintenance. However, they have low resolution and a slow reaction time, they are temperature dependent and if they dry out they need to be uninstalled, re-saturated and installed again.

[5] http://extensionpublications.unl.edu/assets/pdf/ec783.pdf

Heat Dissipation Sensors

These sensors are based on the fact that dry materials heat up faster than wet materials because of heat dissipation produced by the thermal conductivity of water. Therefore, increased water content in a porous material increases in proportion to heat flow. A thermal heat probe has a porous block combined with a heat source and an accurate temperature sensor. The heat source works for a few seconds and the temperature sensor measures the temperature before and after heating to calculate the difference. These sensors are sold with the calibrated relation between the measured change in temperature and soil water potential.

These sensors have a wide measurement range, need no maintenance, have a 10-cm measurement cylinder radius, can give continuous reading and are not affected by salinity because measurements are based on thermal conductivity. However, they need a sophisticated controller or data logger to control heating and measurement operations, have a slow reaction time (do not work well in sandy soil where water drains more quickly than the instrument can equilibrate to) and have fairly large power consumption for frequent readings.

2.2.4 *Machines and Their Properties*

2.2.4.1 Travel Speed Sensor

This sensor determines the distance the tractor or combine harvester travels. Sometimes travel speed is measured with a GNSS receiver or a radar or ultrasonic sensor. An average radar or ultrasonic sensor speed sensor has an accuracy $<\pm5\%$ for speeds up to 3 km hour^{-1} and $<\pm3\%$ for larger speeds.[6]

2.2.4.2 Tractor Sensing Systems with ISOBUS

From the early 1980s with the development of microcomputers, the first attempts to record tractor performance data started by measuring draft forces, velocity, fuel consumption, engine load and wheel slip values (Harter and Kaufman 1979; Grevis-James et al. 1983). The technological innovations of on-board tractor performance monitoring systems and the recent advances in tractor technology enable the acquisition of tractor and implementation status. This has been achieved through the agricultural machinery industry protocols SAE J1939 (Society of Automotive Engineers 1995) and ISO 11783 (or ISOBUS) and provide useful information to optimize overall field productivity (Scarlett 2001; Backman et al. 2013). Combined with the GNSS, the system could be used for spatial mapping of tractor–implement field performances (Taylor et al. 2002; Yahya et al. 2009). Moreover, the ability to monitor and collect tractor and implement performance data can benefit management decisions and lead to fuel savings (Tsiropoulos et al. 2015).

---

[6] http://www.dickey-john.com/product/radar-ii/

Information-to-action decision-making processes, as well as precision agriculture applications, require sensors for on-the-go data collection of crop and soil variation(e.g. soil moisture content, NDVI, crop density, and so on). The ISOBUS protocol can play an important role in the development of precision agriculture and helps information to be exchanged and stored more efficiently between sensors, processors, controllers and software packages from different manufacturers within the same tractor or vehicle (Stafford 2000).

From 2001 on, the ISOBUS standard matured and became the international standard that was adopted by the agricultural machinery industry. Tens of thousands of ISOBUS implements, tractors and components were sold successfully worldwide, but despite this large number there were also 'incompatibility problems' to be solved. Farmers or contractors that purchased equipment based on this standard were often promised that the investment in ISOBUS was a secure investment and would give a "Plug and Play" solution for all their needs. However, after a few years in practice this promise appeared to be very different. While the industry, together with DLG (the only ISOBUS test entity in the world), focused on the technical aspects for testing and certifying components, the practical implementations in the field appeared to be problematic sometimes leading to situations where the end-customer simply did not have a working solution between cross-branded equipment (Vlugt 2013).

Another basic problem is the challenge to integrate the data of these new technologies into a coherent farm management system. The main problem arises from the heterogeneous nature of these data resulting in a variety of data formats and interfaces. Incompatibility of different data formats is usually a fundamental problem and considerable manual effort is required just to convert data from one format to another. Therefore, there is an imperative need for continuous data exchange, either between the farm's computer and the computing devices mounted on the farm machinery or between the farm's computer and the external farming systems such as contractors, suppliers and advisory services, and so on. A research team at Iowa State University has developed a data logging platform (CyCAN), a standalone ECU aimed specifically at quantifying the key properties of agricultural machinery (Darr 2012). The CyCAN data logger connects directly to the ISOBUS port in the tractor cab and provides direct access to all available CAN Bus information. Steinberger et al. (2009) presented a prototype implementation of an agricultural process-data service that enables flexible data networking based on the farming standard without much complexity for the farmers or farm managers. The data are recorded through the ISOBUS port and transferred to a server where data are analysed and aggregated to completed jobs and can be requested for further use by a web portal and a web service interface. Tsiropoulos et al. (2013) presented a management information system for spatial analysis of tractor-implement draft forces. The system can record and combine data in real-time from ISOBUS, CAN Bus and various types of analog and digital sensors (fuel meters, load-cells, etc.). The data can be transferred in real time to the system database and the results are analysed spatially. The system was expanded (Gravalos et al. 2014), and the measurements with soil moisture sensors were transferred to the management system with remote terminal units and a wireless gateway installed into tractor cabin.

In farming businesses, however, data exchange requirements are not fixed and changes occur frequently. The data exchange techniques usually lack flexibility with regard to efficient management of required changes, and the system often needs manual maintenance. The low-level hand held data conversion from one format to another usually requires a lot of manual work, which causes problems and is confusing for ordinary farmers. Iftikhar and Pedersen (2011) proposed an easy-to-use and flexible solution for ISOBUS based bi-directional data exchange as well as efficient requirements weth changes in management. The system uses an XML-based graphical user interface that generates high-level data exchange specifications that can be used by farmers or farm managers. The solution is expected to work well in low-bandwidth and partially disconnected environments, and where the data exchange requirements are not fixed and changes occur frequently, as in the farming business. The authors also pointed out the future need to implement a rule-based tool for bi-directional exchange of data that will provide the underlying rules of an interactive procedure for generating high-level data exchange specifications with ease-of-use.

2.2.4.3 Unmanned Aerial Vehicles (UAVs)

An unmanned aerial vehicle (UAV), commonly known as a "drone" is an aircraft without a human pilot aboard. The flight of UAVs may be controlled either autonomously by on-board computers or by the remote control of a pilot on the ground or in another vehicle. There are two main platforms for UAVs: fixed wing and multi-rotor.

A fixed wing platform has the advantage of covering large areas efficiently, whereas a multirotor is able to remain very stable in challenging conditions with large loads. The UAVs are equipped with a GNSS receiver that is used primarily for location information for the autopilot and of course for the data recorded to be linked to its spatial position. In addition, UAVs have autopilots in order to be programmed to fly over a certain area and record the desired data. In many cases, UAVs communicate with a ground control station (GCS) by radio link. The GCS is usually just a laptop computer with software such as Mission Planner. The same software is also used to set the flight paths for the UAV missions. Many UAVs are equipped with a u-blox GNSS receiver or similar, which is compact and provides 1 m and 2 m vertical and horizontal accuracy, respectively.[7] These receivers also include an inertial measurement unit (IMU) for detecting changes in pitch, roll and yaw and for enabling dead reckoning capabilities. These systems are affordable and are accurate for most situations.

The UAVs already offer new alternatives for agriculture and other applications in which high spatial resolution imagery delivered in near-real time is needed (Herwitz et al. 2004). Diagnostic information derived from images recorded from on-board sensors such as biomass, leaf area index (LAI), disease and water stress can thus inform decision-making in crop management, yield forecasting and environmental protection (Zhang and Kovacs 2012). When imaging sensors are used with UAVs,

[7] http://www.dji.com

overlapping images are required to achieve full cover of the field under investigation to produce an ortho-mosaiced image.

Comparison with manned aircraft:

- UAVs can be flown in dangerous situations (no pilot or scientist on board).
- UAVs can fly for long durations, on dull missions such as mapping or for diurnal measurements without inconveniencing pilot or crew.
- UAVs with long endurance can remain still during an emergency, enabling long-term awareness of a situation.
- UAVs with a long range capability can be launched from or flown to a remote location.
- UAVs with high altitude capability can fly safely above theweather and above air traffic.

Comparison with satellites:

- UAV pictures are not disturbed by clouds because their flying height is low.
- UAVs can fly to precisely selected locations at precisely selected times.
- UAVs can be tasked to remain over arbitrary targets for long durations.
- UAVs can carry a variety of interchangeable high resolution imaging instruments.
- UAVs are recoverable for maintenance and upgrades of sensor and communication systems.

2.2.4.4 Unmanned Ground Vehicles (UGVs)

Basic characteristics for prototype robots are their light weight, small size and energetic autonomy (Blackmore 2007). Light weight means that the vehicle requires less energy and induces less soil compaction, and they must be small for safety reasons, to achieve greater precision on their tasks and to have more manoeuvrability.

Mechanical design of the prototypes depends on its main tasks or developers' goals. The UGVs can run on tracks or wheels. Even though tracks have many advantages compared to wheels with zero turn radius, better flotation, smoother ride on rough surfaces, greater efficiency over a wider range of soil conditions and more stability on hillsides, their main disadvantages for use on robots are the motion control and the pose estimation (Martínez et al. 2005). Most researchers use wheels on their prototypes because tracks require the use of complex dynamics (or effective kinematics approximation), combined with their higher price and higher cost of maintenance. To achieve maximum manoeuvrability, which is very important for autonomous vehicles, 4-wheel drive and steering (4WD/4WS) is commonly used (*BoniRob*, Ruckelshausen et al. 2009, *HortiBot*, Jorgensen et al. 2007, *Roboturk*, Tekin et al. 2013, *Zeus*, Tressos 2011, *API*, Danish Technical University 2006, *AgRover*, Tu 2013). To be able to work with different types of crop and for maximum flexibility, many prototypes have variable track width and height configuration (Bonirob, Zeus, API, AgRover). This is achieved with the use of fixed wheels on arms or legs that can move separately. The problem with this type of approach is the vehicle's centre of gravity, especially when the height increases, which make prototypes unstable on slopes. To reduce stability problems on slopes, HortiBot and Roboturk have very low centres of gravity; HortiBot can work on slopes up to 40°. One other interesting approach for working on slopes is AgRover, which has a self-levelling pneumatic system for maintaining the platform flat and stable. In addition, there are some totally different approaches based on needs such as the Agricultural Mobile Robot (Tabile et al. 2011), which was constructed to operate in crops up to 1.8 m in height with variable intra-row spacing. Most of prototype chassis are made from steel for greater durability (BoniRob, HortiBot, Roboturk, Zeus) or aluminum (API, AgRover, SlugBot) for less weight. Almost all prototypes are lightweight starting from 100 kg (*Dionysus*, Hau and Cereteth 2013) with very few prototypes over 400 kg (e.g. Roboturk).

Power sources that are commonly used on prototypes are petrol engines (HortiBot, Agrobot, Dionysus) or electric motors (BoniRob, Roboturk, Zeus, AgRover). Electric motors are environmentally friendly, but petrol engines have more power. For that reason the selection of power source depends on the tasks, use and design. Some researchers have created hybrid prototypes (Halmstad weeding robot, Åstrand and Baerveldt 2002), but generally this approach is being rejected because of the increase in total prototype weight. A unique and interesting project was the SlugBot project (Kelly and Melhuish 2001) which aimed to create a robot predator developed to investigate issues of energy autonomy, by harvesting slugs and putting them into a digester to power the robot. From all agricultural attempts to create robots, it was shown that energy efficiency can be improved by constructing the robots with light but strong materials such as carbon fibre and aluminium, and by using decentralised modern low power controllers and electronics where possible (instead of a single high speed central processor).

2.2.5 Data Analysis and Evaluation Technologies

These SFTs are used for analysis of the data obtained from the data acquisition SFTs and are categorised as follows:

- Management zone delineation
- Decision-support systems (DSS)
- Farm management information systems (FMIS)

2.2.5.1 Management Zone Delineation

All data collected have to be analysed and interpreted if a meaning is to be drawn from them. There are generally too many data and appropriate methods that exist or have to be developed for the analysis need to be applied. Simple exploratory (descriptive) statistics can give a first impression of the values, their spread, the range and the distribution. Geostatistics, based on what is called 'the theory of regionalised variables', is basically a probabilistic method of spatial interpolation. Final construction of the map at the local level is made possible from estimated values based on the variogram by kriging; the variogram describes the structure of the spatial variation of the sampled data. This type of information, which can be obtained for different properties and for successive years, opens new and interesting possibilities in agronomic crop analysis and management (Arnó et al. 2009).

Variograms are used to assess the spatial variation of the measured values. For each property semivariances are plotted against the distance (lag) between the points. A model is fitted to the experimental variogram, which is the theoretical variogram. Maps covering the whole field can be produced and indicate the variation in the properties. There are several methods of data analysis, although that there is not a clear way to compare the maps produced. This is still based on an optical impression for comparison of the maps. Correlations between parts of the field with different peoperties can be carried out to assess their relations. Kitchen et al. (2005) tried to delineate productivity management zones based on soil electrical conductivity (ECa), elevation and yield maps by management zone analysis (MZA). They used the agreement between pixel in zones to compare the zones based on different variables. Tagarakis et al. (2011) have applied this methodology to precision viticulture with encouraging results.

The analysis of the data aims to define parts of the field with common characteristics that can be managed separately. These are the management zones. Delineation of management zones should create homogeneous parts of the field where inputs or other practices can be applied in the same way. The management zones should be large enough to permit variable-rate application of inputs, but small enough to be homogeneous.

2.2.5.2 Decision-Support Systems

A decision-support system (DSS) is a computer-based system that supports business decisions. In agriculture it refers to the decisions taken by the farmer for farm management. Precision agriculture is connected directly to decision making by the farmer. It can be described as an example of the conversion of data into decisions (McBratney et al. 2005). It is quite true that research has not been successful in developing DSS at the moment. The lack of functional tools for decision-taking explains, to certain extent, the difficulty for a rapid and widespread adoption of PA. Arnó et al. (2009) pointed out that the development of DSS in PA undoubtedly remains a pending assignment. Kitchen et al. (2005) indicated that more precise crop models in PA might help in the development of successful DSS. The inadequate development of control and decision support systems for implementing PA decisions has been identified as a major stumbling block to the adoption of PA (McBratney et al. 2005).

2.2.5.3 Farm Management Information Systems (FMIS)

Agriculture has become very complex and farmers using Smart Farming Technologies (SFTs) acquire a vast amount of data that have to analyse and derive the best decisions for their crop management. The key to success is access to timely information and elaborated decision making. Decision making is an important aspect in farm management and has been studied by numerous authors and with different applications (i.e. Sørensen 1999; Fountas et al. 2006). Farm Management Information Systems (FMIS) is defined as a planned system for collecting, processing, storing, and disseminating data in the form needed to carry out farm operations and functions (Sørensen et al. 2010). The fundamental components of FMIS include specific farmer-oriented designs, dedicated user interfaces, automated data processing functions, expert knowledge and standardized data communication and scalability. To improve functionality, various management systems, database network structures and software architectures have been proposed, where FMIS have increased in sophistication through the integration of new technologies, such as web-based applications and applications for smart phones and tablets (Nikkila et al. 2010). As agriculture is a complex system it incorporates a number of interactions between farmers, advisors, traders, sellers, governmental bodies, farm machinery, environmental regulations, economic estimations and others. FMIS can cover a large number of functions, such as inventory, calendar, direct sales, site-specific management functions. A set of 10 functions are presented (Table 2.3), modified by the set of functions proposed by Fountas et al. (2015).

There is a large discrepancy between the functions that are provided by research FMIS and their commercial counterparts. Fountas et al. (2015) compared the published FMIS from academic institutes and 141 commercial software and they concluded that academic research tends to analyse more complex systems, capturing new trends involving spatial and temporal management, distributed systems involving internet of things, future internet and web services. Commercial applications tend to focus on solving daily farm tasks with the aim to generate income for the farmers through better resource management and field operations planning. The advances that are needed in the development of FMIS include improvements in technology, adaptation motives, specific new functionalities and greater emphasis on software design governed by usability and human–computer interaction. The diffusion of information management as business innovation in the farming community could benefit from the comprehensive research developed in the last decades on the adoption of ICT and e-commerce among both consumers and small businesses.

The profitability and adoption of FMIS cannot be estimated easily and there is limited research or commercial research available to support this. An adoption study by Lawson et al. (2011) in four EU countries (Denmark, Germany, Finland, Greece) revealed that the benefits from reduced labour costs of introducing advanced FMIS could be related to budgeting procedures, field planning and paperwork for subsidy applications and public authorities. Northern European farmers are inclined to spend more time working with computers than their Southern colleagues, probably because of the more developed and more business-oriented types of farms that exist in Northern Europe. About 30% German, less than 20% Danish and over 20% Finish respondents were positive about the use of computers for documentation when dealing with government agencies. In Germany, the majority of the respondents spent 20 hours per week on inside-office tasks (i.e. time at the computer, for preparation of applications for area subsidy etc.) as well as learning new procedures etc. The 20 weekly hours spent by German respondents is significantly different from the 7, 3 and 1 hours spent, respectively, for Denmark, Finland and Greece. The average of 25% of the inside-office time allocated to field management planning by the German respondents is considerable larger than the 14, 15 and 16% allocated by Greek, Finnish and Danish farmers, respectively. The 3% of the time budget allocated to private tasks in the office by the Greeks is significantly less that of the 14, 11 and 8, respectively, for the Danish, Finnish and German respondents. In Northern European countries, spring time is the most intensive when farmers finalize field plans and fill complex subsidy applications at the same time.

The FMIS could be classified according to the application. There is software available for crop monitoring, for whole-farm management, for precision agriculture only and specialized software for specific applications (scheduling irrigation, spraying prognosis, accurate weather forecasts).

2.2.5.4 Software for Whole Farm Management, Forecasting and Crop Monitoring

In Table 2.4 there are examples of software offered for different purposes. Some software is multi-purpose focusing on whole-farm management in the fields, for precision agriculture applications, inventories, sales, planning and reporting for the single payment scheme. Some other software is dedicated to specific applications, such as for forecasting specific diseases or pests, for irrigation only or for pasture management.

We have to note that these are some examples of software packages in the market. There are many ‘start-up’ companies that have developed their own software and these also operate as international companies. The purpose of Table 2.4 is to demonstrate the variation in software available in the market and their functions. For more information on clustering of software in agriculture, a review study can be found at Fountas et al. (2015).

2.2.6 Precision Application Technologies

This category of SFTs refers to the technologies that are used to apply the decisions taken (using data analysis & evaluation technologies) after receiving the information on the condition of the farm (using data acquisition technologies). They are divided in:

- Variable-rate application technologies
- Precision irrigation
- Precision weeding
- Machine Guidance

Table 2.3 Farm management Information systems functions

Function title	Function description
Field operations management	Recording of farm activities to help farmer optimize crop production by planning activities and observing the actual execution of planned tasks. Preventive measures may be initiated based on the monitored data.
Best practice (including yield estimation)	Production tasks and methods related to applying best practices according to agricultural standards (e.g. organic standards, integrated crop management (ICM)). A yield estimate is feasible through the comparison of actual demands and alternative possibilities, given hypothetical scenarios of best practices.
Finance	Estimation of the cost of every farm activity, input–outputs calculations, equipment charge-outs, labour requirements per unit area. Projected and actual costs are also compared and input into the final evaluation of the farm's economic viability.
Inventory	Monitoring and management of all production materials, equipment, chemicals, fertilizers, and seeding and planting materials. The quantities are adjusted according to the farmer's plans and customer orders.
Traceability	Crop recall, using an ID labelling system to control the produce of each production section, including use of inputs, employees, and equipment, which can be easily archived for rapid recall.
Reporting	Creation of farming reports, such as planning and management, work progress, work sheets and instructions, orders purchases cost reporting, and plant information.
Site Specific	Mapping the features of the field, analysis of the collected data, generation of variable rate inputs to optimize input and increase output. This is the SFT component. It could be a separate software or could be integrated.
Sales	Management of orders, charges for services, online sales.
Machinery management	Includes the details of equipment usage, the average cost per work-hour or per unit area. It also includes fleet management and logistics.
Human resource management	Employee management, availability of employees in time and space, handling work times, payment, qualifications, training, performance, and expertise.

Modified by Fountas et al. (2015)

2.2.6.1 Variable-Rate (VR) Application Technologies

Variable-Rate Granular Fertilizer Application

Variable-rate application of fertilizer implies that the mass flow rate and subsequently the application rate of fertilizer needs to be varied within the field. The spinner and pneumatic spreader are generally the most used of fertilizer application machinery, also fertilizer drills are used frequently. In general, the application rate is changed by changing the mass flow of fertilizer to the delivery system of the spreader (spinning disks or air boom). Current technology allows the rate between different swaths and in the longitudinal direction within one swath to be changed.

The spinner spreader, also called centrifugal spreader, is the most commonly used. Particles fall on a spinning disk which is equipped with vanes and throws the particles into the field. Variable-rate control systems generally change the mass flow rate from the hopper to the delivery system. In the case of the centrifugal spreader, this is generally done in two ways: by changing the size of the orifice at the bottom of the hopper (Chen and Shiping 2011), or by changing the speed of the conveyor belt (Akdemir et al. 2007; Fulton et al. 2001) or metering rollers (Behic and Okyay Sındır 2013) that deliver fertilizer to the spinning disks. Some systems use load cells to measure the dynamic (measured around 100 Hz) weight of the spreader with fertilizer, based on this, they predict the flow rate and provide online feedback. A second reference sensor (often an accelerometer) is then used to compensate for varying field conditions (Van Liedekerke et al. 2006). Measurement of the mass flow is also possible by measuring the torque to rotate the spreading disks. This allows differentiation of fertilizer quantities spread from the left and right spreading disk.

Pneumatic spreaders use airflow to convey fertilizer particles from the metering units to distributors on the spreading boom. In contrast to the spinner spreader, material is distributed uniformly through the distributors along the length of the boom. Therefore, no overlap is necessary between subsequent swaths.

Fertilizer drills can be used to aim for more accuracy placement (Maleki et al. 2008a) and can be mounted on a row-crop planter (Maleki et al. 2008b). In contrast to the above mentioned broadcast spreaders, the width of the machine equals the working width and particles are not thrown into the air which reduces the sensitivity of the application system (e.g. for wind). Based on the ground speed of the tractor and the prescription map or online sensor values (Maleki et al. 2008a), the mass flow rate of fertilizer is changed by controlling a metering screw (Forouzanmehr and Loghavi 2012) or an electrical actuator changing the rotational speed of the fertilizer metering devices (Maleki et al. 2008a). Important factors that increase the accuracy of the VR system are the different delay times that occur in the process. If an online sensor is installed to determine the application rate, the acquisition and processing time of the data should be taken into account. Secondly, changing the position of a metering screw or actuator takes time. Another source of lag time is the time required to reach the end of the fertilizer tube (positioned at the furrow openers) after setting the new mass flow rate at the metering device.

Variable-Rate Lime Application

Under application of lime can cause large yield losses. However, its over application can be as detrimental as under-liming because it is costly and can create problems with availability of some nutrients (e.g. inhibits P and Zn or leads to toxic concentrations of available Mn), disease pressure, reduced herbicide performance and herbicide degradation (Weisz et al. 2003; Kuang et al. 2014). Over- and under-liming cannot be avoided if lime is applied uniformly throughout the field. Variable-rate (VR) lime (primarily $CaCO_3$) application can increase crop yields and the economic return of the farm (Weisz et al. 2003). Lime application increases the soil's pH to a desired level and an optimal pH level in the soil is important to achieve optimum yields and consistent quality (Kuang et al. 2014). Also, lime improves the uptake and availability of plant nutrients and can also improve water penetration.

Both spinner type spreaders and pneumatic applicators exist for lime applications. In spinner spreaders a conveyer belt or chain transfers the materials from the

Table 2.4 Software packages offered in agriculture

Software	Description
FARMSTAR	FARMSTAR is a satellite technology-based service devised and delivered by Airbus Defence and Space since 2003. FARMSTAR's users are taking advice on precision agro-management knowing the exact time and area where they should apply **fertilizer and pesticides**. Satellites flying over the fields take accurate measurements of the radiant solar energy absorbed and reflected from the surface across the farm terrain. The value of the reflected energy varies according to the level of growth of the vegetation, thus satellite measurements can indicate crucial field factors such as soil moisture, surface temperature, leaf cover and level of chlorophyll. Personalised "recommendation cards" divided into very small areas of the field are provided to each user, offering her/him prescriptions for the necessary amounts of chemicals that should be applied, as well as where and when to be applied. The average price for FARMSTAR services to the farmer **is €10/ha**. By 2011, coverage was 440,000 hectares and 10,000 farmers had subscribed to it, while the number of hectares has already reached to 1 million hectares this year (personal communication).
Akkerweb	Akkerweb is Dutch software used for **whole-farm management** of the fields with **precision agriculture** capabilities. It has a number of applications such as herbicide application, nitrogen fertilisation and pesticides applications. So far, it has about 25,000 fields registered in their database.
365farmnet	365farmnet is a **whole-farm management software**, which is sold to farmers to optimize their production including inventory, sales, precision agriculture functions and connections to sensors.
Farmworks	Farmworks is a Trimble company that offers **whole-farm management** including inventory, sales, planning and all **precision agriculture** functions for VR seeding, fertilization and spraying. It offers different packages to contractors and single farmers.
John-Deere Farmnet	John-Deere Farmnet is a **whole-farm management** system to provide **precision agriculture** functions together with other farm management services. It is linked to John Deere equipment and generates site-specific maps based on field data.
SST-toolbox	SST-toolbox is a USA-based company that offers **whole-farm management** software focusing on **precision agriculture** applications, having different packages for contractors and single farmers.
PIXAGRI	TerraNIS' PIXAGRI is a decision support tool for farmers based on remotely sensed data which allows detection of sub- and inter-field variations and is designed to optimise the application of agricultural inputs. It is a generic product suite based on optical satellite imagery allowing farmers to **identify and control the agronomic factors limiting their yield**. Aimed primarily at farmer's cooperatives, PIXAGRI is available for all types of crops and most suitable to large areas (+2000 ha). The product suite is comprised of agriculture maps over the territory, analysis of yielding capacities in a region and heterogeneity and vigour maps. It is sold as a service that delivers maps providing crop status information and is commercialized through a yearly subscription over a fixed territory. The price is **5–10 €/ha/year**, depending on the level of information required. Currently, TerraNIS commercialises the PIXAGRI product range in Midi-Pyrenees and other regions in France, Canada, Hungary, Serbia (via Airbus Defense and Space).

(continued)

Table 2.4 (continued)

Software	Description
SOYLsense	SOYLsense uses satellite imagery to measure crop canopy variations. Leaf Area Index (LAI) maps with information of the requirements of **nitrogen application** are produced based on the data obtained, allowing the optimization of nitrogen rates. SOYLsense enables farmers to monitor their field and acquire advices about the application of nitrogen fertilizers. Moreover, users can view their LAI map online and have the ability to create and process their own nitrogen application maps.
FarmRite	BlackBridge's RapidEye constellation of satellites in cooperation with SST Software, a global provider of GIS software for agriculture, delivered the SST FarmRite service offering farmers and agronomists the appropriate data in order to generate **nutrient and pesticide application maps on their own**. Users can order products and acquire the desire end products, while management reports are provided to users, allowing them to track the success of product offerings and make the appropriate adjustments.
FarmingTruth	The FarmingTruth project -funded by ESA- aimed to deliver a precision agriculture service that enables users to collect and utilize crop and soil data from various data sources including Earth Observation and Satellite Navigation. Among FarmingTruth providing data are recommendations about variable rate **fertilizer and lime application**.
Ag Data Viewer	Ag Data Viewer is a precision farming software package that provides a wide range of precision agriculture practices from satellite imagery. Among others it offers variable rate application maps with recommendations about **fertilizer inputs**.
TalkingFields	Started as an ESA-led project, TalkingFields is an operational cost-effective Precision Farming service, combining GNSS technologies with Earth Observation (EO). For a terrain located using Navigation Technologies, the user requests a specific service from the TalkingField's catalogue, such as the "Improved Soil Mapping", the "Economic Evaluation" or the "Yield Estimation". Based on navigation information, EO measurements and land surface modelling, the provider prepares the **custom made service and offers users recommendations for individual farm treatment**, application of cost-effective practices and more effective farming systems.
HydroBio	HydroBio provides farmers with an **irrigation decision support system using weather and Earth Observation data** to estimate the precise water needs of each field along with crop monitoring. HydroBio users receive information that enables them to deliver an optimal irrigation strategy to their field including maps for the previous water use, recommended irrigation schedules, strategies to meet irrigation deficit when necessary, information about the best applied practices and what needs to be improved.
SPIDERwebGIS	SPIDERwebGIS is an Information System, open source, based on Web 2.0 designed as a participatory tool to aid decision making applied to improve **agroforestry environments** at different levels of management and whose main source of information are products obtained from time series of Earth observation satellite images.

(continued)

Table 2.4 (continued)

Software	Description
PiMapping	eLEAF's PiMapping (Pixel Intelligence Mapping) Technology uses satellite imagery, weather information and precipitation data to create "smart pixel" images providing information about cropped surfaces and enabling the development of applications for the agricultural sector. It provides farmers with information such as growth, moisture and yield for almost all cropped surfaces, and offers them the ability to better manage their **irrigation practices** by delivering daily data about their crop's actual evapotranspiration, biomass production and crop water requirements.
Synfield	Synfield is a system that utilizes meteorological data for estimation of evapotranspiration and together with soil moisture sensors it estimates the **irrigation needs**. It has developed a web interface to estimate the water needs and installs electro-valves in the field to remotely activate irrigation.
CropLook	CropLook inform farmers with potatoes and wheat crops about their crop growth in a weekly basis, offering them growth parameters such as **crop evaporation, nitrogen content, and yield figures**. CropLook using remote sensing, the satellite information are processed with the ETLook algorithm to be translated into crop data, and this data are provided directly to the user via a web portal.
GrapeLook	The GrapeLook project, based on satellite technologies (earth observation, satellite communication and navigation) as well as terrestrial technologies, aimed to help **grape farmers with the irrigation water resources and nitrogen applications in the Western Cape, South Africa**. All data obtained, such as soil moisture levels; irrigation retrieval from evapotranspiration updates; and digital boundaries of vineyard blocks, were uploaded on the project website, which were publicly accessible. Moreover, farmers received an irrigation forecasting tool and a SMS/MMS service with information on irrigation scheduling and fertilizer application.
FruitLook	FruitLook, a successor of GrapeLook, covers a larger area of crops including deciduous fruit trees. Weekly data on crop's growth, **water use and nitrogen conten**t are available to farmers for free. In the near future, the developers of FruitLook will release an "irrigation planner" which will inform farmers about when, where and how much they should irrigate to avoid water stress in their crops.
The Pastures from Space	The Pastures from Space project aimed to deliver near real-time information tools at a **whole-farm and within-paddock level of forage crop production**. The Pastures from Space provides estimates of pasture production during the growing season by means of remote sensing. Satellite data is used to accurately and quantitatively estimate Pasture Biomass or Feed On Offer (FOO) or combined with climate and soil data is used to produce Pasture Growth Rate (PGR) estimates.
URSULA Agriculture	URSULA Agriculture provides a suite of services based on imagery from drones, aircraft and satellites. Products include Ursula **Crop Performance** (a crop monitoring service which measures within field variation across the growing season), URSULA Scout (a rapid mapping and quantification tool for highlighting areas of crop stress), URSULA Trials (a quantitative comparison tool enabling the evaluation of agricultural trials), URSULA Farm View (aerial imaging tool enabling visual assessment of crop progress) and URSULA Compliance (a service aimed at supporting CAP subsidy claims).

(continued)

Table 2.4 (continued)

Software	Description
Irrisat	Irrisat is a research project that is focused on providing information on **irrigation needs** to the farmers and Irrigation User Associations based on Earth Observation satellites. Except of irrigation related information, it offers crop vigor information along with meteorological data.
Oenoview	Oenoview is an innovative remote sensing tool that produces a cartography of vineyards. This information is used by viticulturists in order to apply **site specific management at their vineyards** and monitor them at sub parcel level.
AGRIVI	AGRIVI is a platform that offers **complete farm management**. It has the ability to monitor weather and provide alerts on crop pests and diseases. It offers tools for crop management, farm economics, resources and inventory, growing analytics and reports.
FARMBRELLA	FARMBRELLA is a software service addressing the real farmers' needs of the meteorological data and analysis. It offers the ability to the farmers to have **hyper local weather informatio**n of their fields while it measures heat sum and provides alerts on extreme crop conditions.

https://www.farmstar-conseil.fr/
http://terranis.fr/en/pixagri/
http://www.soyl.com/index.php/services/soylsense
www.sstsoftware.com/products/farmrite/
www.farmingtruth.com/
http://www.geektechforag.com/
www.talkingfields.org
http://hydrobioars.com/
http://www.spiderwebgis.org/
www.eleaf.com/technology-pimapping
www.synelixis.com
www.croplook.com/
https://artes-apps.esa.int/projects/grapelook
www.fruitlook.co.za/
www.pasturesfromspace.csiro.au/
www.ursula-agriculture.com/

hopper to spinning disks. In variable-rate systems the application rate is controlled by adjusting the gate opening and or changing the speed of the conveyor (and thus the input rate of material). In pneumatic applicators the material is spread by an air stream through a piped boom (Grisso et al. 2011).

Variable Rate Manure Application

There are two different levels for variable-rate slurry application. For the first level, only the application rate, i.e. the flow of slurry from the tank to the application hoses, is varied. However, because manure is not consistent in nutrient content, a second level was designed. Therefore, at the second level, the nutrient content of the slurry should be measured online by sensors that measure electrical conductivity or use NIRS (Calcante et al. 2015).

In general, slurry can be delivered in two ways to the delivery system: (1) by putting the tank under pressure or (2) by pumping the slurry. With a pressurized tank, the application rate can be modified by changing the size of the gate opening that delivers slurry from the tank to the delivery system. Calcante et al. (2015) used a variable gate hydraulic valve, powered by the main hydraulic system of the tractor. Based on the nitrogen content of the slurry (measured before application), the ground speed of the vehicle (measured with a sensor or using GPS information) and working width, the required slurry flowrate can be calculated and set by the controller (Brambilla et al. 2015). Sensors that measure flowrate are used for feedback to the rate control system, such as Doppler effect sensors (Calcante et al. 2015; Brambilla et al. 2015) and electromagnetic flowmeters (Morris et al. 1999). In general, these give a more accurate estimate of the flowrate than load cell measurements (Morris et al. 1999). Funk and Robert (2003) used a pneumatic pinch valve for both flow metering and flow measuring.

In the other case, slurry is pumped from the tank to the applicator by a centrifugal or positive displacement pump controlled by the tractor PTO (Funk and Robert 2003). In most cases, the pump is driven by the tractor PTO. The application rate is varied by changing pump or valve settings based on the online measured flowrate.

Variable Rate Pesticide Application – Map-Based System

The VR pesticide application technologies enable changes in the rate of application to match actual or potential pest stress in the field and avoid application to undesired areas of the field or plant canopies (Karkee et al. 2013). They can also significantly reduce spray overlap (Batte and Ehsani 2006). In general, weeds have received the greatest attention from developers of site-specific technologies because of their immobility (Swinton 2003). The VR technologies for pesticide application can also be used to apply fertilizer at variable rates (Ess et al. 2001).

Map-based VR pesticide application adjusts the application rate based on an electronic map, also called prescription map or application map. Using the field position from a GPS receiver and a prescription map of desired rate, the input concentration is changed as the applicator moves through the field (Grisso et al. 2011). Two main categories can be distinguished, i.e. (i) rate control, including flow-based control systems, direct chemical injection systems and chemical injection systems with carrier control and (ii) nozzle control, including modulated spraying nozzle control systems.

Variable Rate Pesticide Application – Real-Time Sensor Based Spraying

Real-time sensor based spraying controls the application rate based on the current situation of pest stress or canopy characteristics. These systems involve both contact and non-contact sensing to identify either pests that need to be controlled or the crop and foliage or canopy that needs to be protected. Various types of sensors can be used such as colour cameras, photodetectors, laser scanners, multispectral and hyperspectral cameras, thermal cameras, and ultrasonic sensors. These sensors have been used to determine variables such as colour, shape, size, texture, reflectance and temperatures of pests. This information is then used to categorize pest or canopy patterns, and to identify and locate them. The sensor input can also be used to control the direction and rate of chemical application (Karkee et al. 2013). The same rate and nozzle control systems as in map-based VRA can be used.

In addition, sprayers that use information on the environment to reduce drift are currently being developed. These sprayers use, for example, sensors that measure the wind speed and direction and change the sprayer settings (spray pressure, nozzle type) accordingly depending on where the sprayer is in the field in relation to vulnerable areas based on GPS (Doruchowski et al. 2009).

Boom Height Control

Boom oscillations and vibrations are disastrous for the homogeneity of the spray liquid distribution on the crop, resulting in under- and over-applications of chemicals with, respectively, a missed treatment effect and remaining residues (Hostens et al. 2000). From simulations performed by Ramon et al. (1997), it was concluded that both rolling motions and horizontal vibrations of the boom can severely disturb the spray deposition pattern. Local under- and over-applications caused by boom rolling varied between zero and 10 times the desired dose. Horizontal boom vibrations caused variations between 0.3 and 4.0 times the prescribed dose. Varying ground speed, changing tyre pressure and ground unevenness can lead to significant under- or over-application of spray because of oscillation of a sprayer boom above its horizontal axis. Boom height control is used to minimize such effects and improve the uniformity of chemical application (Karkee et al. 2013).

Ultrasonic sensors measure (40 times per second) the distance to the ground. This information allows the control system to make responsive height adjustments. The system has shown reliable control with average speeds more than 29 km hour^{-1} in all kinds of uneven terrain. Although boom height control is not a VRA technology, it eliminates streaks and improper overlaps, and improves coverage (Grisso et al. 2011). Similar control mechanisms can also be used to position the spray tower an appropriate distance from the crop canopy in orchards and ornamental nurseries (Karkee et al. 2013).

Variable Rate Planting and Seeding

The VR planters and seeders can vary the rate of planting and seeding during application. This is often accomplished by disconnecting the planting or seeding system from the ground drive wheel, which usually keeps the planting or seeding rate constant when the speed of the tractor varies. The planting or seeding rate can also be adjusted to the local soil potential by driving the planting or seeding system with an independent engine, gear box (to change speed of the ground wheel input) or hydraulic drive (Grisso et al. 2011).

Most VR planters or seeders will be matched with a prescription map. The VR planting and seeding is useful in very heterogeneous fields (i.e. fields with large wtihin-field variation in water holding capacity or soil organic matter). The VR planting or seeding is ideal for fields with centre pivot irrigation systems, where areas outside the reach of the irrigation system are planted or sown at a reduced rate to avoid water scarcity caused by a too high a plant density (Grisso et al. 2011).

2.2.6.2 Precision Irrigation and Irrigation Scheduling Technology

A number of theoretical and practical foundations can be seen as important in approaching and appreciating a study on irrigation scheduling and associated support functions. Notably these foundations include understanding of:

- water balance in crop production
- commercial irrigation scheduling methods
- remote sensing principles including the electromagnetic interaction with soil and crop media

Better irrigation scheduling can be achieved mainly when water balance components in crop production are known. The most important factors that constitute the water balance equation are[8]:

- Radiation and temperature
- Evaporation from soil surface and crops
- Transpiration from crops
- Land surface water runoff characteristics
- Sub-surface water flow, in and out of the crop location
- Deep soil percolation
- Capillary rise within the soil
- Irrigation (by various methods)
- Rainfall

Research and development into irrigation scheduling methods and systems for irrigation support is a recurring and important theme in PA, with attention to further developments in:

- Soil water status (current)
- FAO (Food and Agricultural Organisation of The United Nations) – method that uses crop coefficient (currently favoured)
- Crop water stress scheduling

Techniques are now emerging that exploit integrated sensor platforms to determine soil moisture and canopy estimates of evapotranspiration as a basis for improving irrigation scheduling and real-time, stress related control techniques for delivery of adaptively-controlled, plant-level irrigation. Water management for irrigation is seen as an essential activity that requires not only considerations for individual farms, but also in a broader context of national assessment needs and as part of the strategic agenda for global monitoring for environment and security (GMES)[9,,10]

Self-propelled irrigation systems consist of centre pivot and lateral move systems that apply water to pasture or crops, generally from above the canopy (Berne

---

[8] See for example the FAO Guidelines on Irrigation and Drainage, FAO Paper 56, Crop Evapotranspiration

[9] www.nereus-regions.eu

[10] www.esa.int/gmes

2015). These systems are most used in irrigation today (e.g. 72% of irrigation systems in USA were sprinkler-based in 2000 according to Colaizzi et al. 2009). In 2009, most self-propelled irrigation systems used mid elevation spray application (MESA), which has an irrigation efficiency of 85%. The latest and future systems consist of low-energy (elevation) precision application (LEPA – low energy pressure application or LESA – low energy spray application), which has a higher irrigation efficiency (97%). These devices (bubblers, sprayers, spinners, and other related spray techniques to apply water) are usually on drop tubes in or just above the crop canopy.

Micro-irrigation is used especially in areas with very scarce water supply. These systems have, compared to sprinkler systems, a greater crop yield, better water use efficiency, less pesticide use because water is emitted at the surface of the desired high value crop, tree or vine and warmer soil temperature (in case of subsurface) (Camp 1998). This system is especially useful in orchards and vineyards or high value crops because of the larger costs.

There are three types:

- Drip and trickle emitters
- micro-sprinkling and microspray
- subsurface irrigation

Although there has been substantial interest in site-specific management, research on spatially variable micro-irrigation systems has been limited. Using pressure or flow sensors, water and dissolved fertilizer applications at each micro-sprinkler could be monitored and controlled. The volume scheduled irrigation strategy and emitter fault diagnosis routines could be made more effective with a differential pressure sensor across each valve to determine individual micro-sprinkler flow rates. Other control strategies would be possible with alternative types of sensors to measure tree water and nutrient demand, and monitor system status.

2.2.6.3 Precision Physical Weeding

The challenge of physical weeding is to obtain a high degree of selective weed control without producing considerable crop damage as a result of weeding (burning, mechanical weed control with knives, discs, hoes, harrows, etc.). Non-chemical weed control methods need to be directed towards a site-specific weeding approach in order to compete with conventional plant protection product applications. Different approaches and prototype systems have been proposed, adjusting the hoeing or harrowing or burning intensity based on the (earlieror real-time) observed soil density or weed density. Precise guidance and detection systems are prerequisites for successful site-specific weed management. Effective detection and identification is a primary obstacle toward commercial development and industry acceptance of robotic weed control machines. The most promising approach for weed detection is a continuous ground-based system that uses image analysis (Martelloni 2014).

2.2.6.4 Machinery Guidance

Auto-Guidance Systems

Auto-guidance systems have gained increasing interest among farmers as they enable farm machinery to follow straight lines to reduce overlaps of the tractor and equipment passes. These systems help farmers to reduce fuel costs, input costs, time, labour, soil compaction and increase the overall field efficiency. Auto-guidance systems are offered as two options, the lightbar and the auto-steer. Both systems use a high accuracy GNSS receiver (Horizontal <1 cm, Vertical <2 cm) to identify the tractor's location in the field (Gan-Mor et al. 2007).

The basic difference between the two systems is that lightbar guidance requires the operator to manually adjust steering, whereas auto-steer technology adjusts the steering automatically, allowing the operator to monitor the field operation of the implement instead of steering. Lightbar technology is offered at a much lower cost and can be easily transferred from one vehicle to another compared to auto-steer technology, which requires higher investment and it can differ from one machinery manufacturer to another.

Guidance systems are regarded as the most adopted PA technologies worldwide and can be used for many field operations such as seeding, tillage, planting, weeding and harvesting (Abidine et al. 2002), and for use with autonomous vehicles (UGVs – unmanned ground vehicles) with the full utilization of the ISOBUS standard ISO 11783 (Backman et al. 2013). It is especially used if the UGV's IMU for enabling dead reckoning capabilities is needed.

Many studies have compared automatic guidance and manually-guided operation. The use of auto-guidance systems on sugar cane planting operations achieved an accuracy of 0.033 m pass-to-pass, which was five times greater than that obtained by the manual steering system (Baio and Moratelli 2011). Rojo and Fabio (2012) evaluated the accuracy, the cane loss and operational field efficiency achieved by an auto-guidance system used to guide a sugar cane harvester over the field when compared to a manually-guided machine. They showed that the use of an auto-guidance system during the day and night periods increased the field pass-to-pass accuracy relative to the planned row track, whereas it did not significantly decrease the sugar cane loss, once the crop was well cultivated. Shinners et al. (2012) studied the effect of driving experience and operating speeds with manually and automatically guided mowers in a variety of field conditions on 15 farms. They estimated an overlap range between 0.4 and 16.13% of machine cutting width with an average loss at 5% of cutting width. They concluded that auto-guidance has been purported to improve efficiency by eliminating time spent covering ground already mowed, reducing operator fatigue and ensuring a uniform cutting pattern and swath density, and showed that auto-guidance systems to steer the mower reduced overlap loss from 5.03% to 2.34%.

Auto-steer reduces the overlap of multiple passes with the tractor, which is caused mainly by operator error or fatigue. The ability to increase speeds during headland turns and more quickly identify re-entry points were recorded to reduce machinery time requirements by 5% for planting and 10% for fertilizer application (Shockley et al. 2011). An RTK-based guidance system was tested for location mapping of planting events occurring on a tractor-drawn tomato transplanter that can automatically create centimetre-accuracy plant maps for subsequent precision plant specific treatment systems (Perez-Ruiz et al. 2013).

Controlled Traffic Farming (CTF)
The CTF is a system that confines all machinery loads to the least possible area of permanent traffic lanes. Current farming systems allow machines to run at random over the land, compacting around 75% of the area within one season and at least the whole area by the second season. A proper CTF system on the other hand can reduce tracking to just 15% and this is always in the same place. The permanent traffic lanes are usually parallel to each other and this is the most efficient way of achieving CTF, but the definition does not preclude tracking at an angle. The permanent traffic lanes may be cropped or non-cropped depending on a wide range of variables and local constraints. Techniques like CTF have the capacity to benefit all types of crop farming. The CTF also allows optimised driving patterns, more efficient operations (i.e. reduced overlaps). As all operations are aligned, input applications can be targeted very precisely relative to the crop rows.

The CTF management can play a key role in sustaining soils and future crop production that are threatened currently by heavy machinery traffic and intensive production systems. To play this role in sustainable intensification, CTF needs to be developed to become a mainstream technology rather than continuing as a niche practice. Therefore, there is a need to facilitate and support the development and mainstreaming of CTF at a time where development in allied technologies such as headland management systems are increasing growers' openness to the adoption of these systems.

When CTF is combined with headland management type systems it further alleviates soil compaction from continually increasing machine weight, and is of paramount importance for EU farmers (estimated approximately 33 Million ha compacted) in terms of yield loss, reduced nutrient and water efficiency, soil degradation and alleviation costs. While management practices such as deep soil loosening, use of certain cover crops and crop rotations can help alleviate some of the damage to soil structure; these approaches are costly and at best only partly successful. Preventing or avoiding damage to soil structure is preferable to alleviation. The CTF offers scope to restrict the extent of soil structure damage. It involves the configuration and application of field or machinery operations in a way that minimises the soil compaction by using permanent traffic tracks. The CTF also enables other compaction minimising traffic patterns, such as load determined traffic routing.

2.3 Smart Farming Technologies Taxonomy

The SFTs can be classified according to the following parameters:

- Farming system type, which is divided into three major farming systems:
  - **Organic farming**: The farming system that relies on biological control and mechanical cultivation to maintain soil productivity and pest control.
  - **Extensive farming**: Low energy input farming that uses small inputs of labour, fertilizer and capital relative to the land being farmed and where conventional practices are carried out.

- **Integrated farming**: The farming system that use good agricultural practices and or integrated crop management strategies to produce high quality and certified agricultural products.

- Cropping systems type, which can be divided into the main clusters that correspond to main cropping environment within EU-28 and have distinctive differences among themselves:
  - **Arable crops**
  - **Forage crops**
  - **Orchards**
  - **Vineyards**
  - **Field vegetables**
- Availability of the technique that is divided into three categories.
  - **Now available**: Precision Agricultural Technologies (PAT)s that are commercially available to be used by farmers today.
  - **Next 5 years**: The availability in the next 5 years is seen as PATs currently under development or at prototype stage.
  - **In the future** (>10 years): PATs that are at the experimental stage in the laboratories or research institutes, such as robotic harvesters, robotic hoeing, etc.
- Level of investment cost that gives the cost involvement of farmers in SFT adoption and is divided into three levels.
  - **Low**: A low cost investment (e.g. parallel guidance with light bars, yield mapping or soil mapping).
  - **Medium**: A medium cost investment (e.g. on-line sensors combined with direct controlling, on-board computers or terminals, parallel guidance with terminals, Variable-rate application of nitrogen).
  - **High**: A high cost investment (e.g. fully applicable PA software, variable-rate applications in many operations, automated guidance system).
- Farmers' motives that correspond to the reasons for farmers to adopt these technologies.
  - **Operational excellence**
  - **License to operate**
  - **Improving the whole-farm information anagement**.

In the following table, the reader can see the taxonomy of all SFTs according to the above mentioned criteria (Table 2.5).

2.4 Smart Farming Technologies Economic Impact

The use of SFTs can provide to the farmers economic benefits that can be seen in Table 2.6.

Table 2.5 Smart farming technologies taxonomy

Main categories	SFT	Farming system	Cropping system	Availability	Investment	Farmers' motives
GNSS Technologies	Global aavigation satellite systems (GNSS)	All	All	Now available	Low to medium	Improving the whole-farm information management
	Differential GNSS					
	Real time kinematic (RTK)					
	Network RTK (NRTK)			Available in 5 years (Galilleo)		
	Wide area RTK (WARTK)					
	Undifferenced GNSS					
	Precise point positioning (PPP)					
	Fast PPP (FPPP)					
Mapping technologies	Elevation maps	All	All	Now available	Low	Improving the whole-farm Information management
	Soil mapping			Now available (some on-the-go sensors are on prototype stage)	Medium	
	Yield mapping			Now available	Low	
	Yield monitor display					
Data acquisition of environmental parameters technologies (Camera based imaging)	RGB cameras	All	All	Now available (Prototypes are also available for reduced size)	Low to medium (depending on the device)	Improving the ahole-farm Information management
	LiDAR sensors					
	ToF (IR) cameras					
	Light curtains					
	Multi/Hyper-spectral cameras					
	Thermal cameras					
Data acquisition of environmental parameters technologies (NDVI measurement)	Spectral sensors	All	All	Now available (prototypes are also available for reduced size)	Low to medium (depending on the device)	Improving the whole-farm information management
	Fluorescence sensors					

(continued)

Table 2.5 (continued)

Main categories	SFT	Farming system	Cropping system	Availability	Investment	Farmers' motives
Data acquisition of environmental parameters technologies (Soil moisture sensors)	Frequency domain reflectometry	All	All	Now available	Low	Operational Excellence/Improving the Whole farm information management
	Time domain reflectometry					
	Amplitude domain reflectometry					
	Phase transmission					
	Time domain transmission					
	Tensiometers					
	Gipsum blocks					
	Granular matrix sensors					
	Heat dissipation sensors					
Machines and their parameters	Travel speed sensor	All	All	Now available/available in the next 5 years (some technologies are already well established others are still under development).	Medium to high	Improving the whole farm information management
	Tractor sensing systems using ISOBUS					
	Unmanned aerial vehicles (UAVs)					
	Unmanned ground vehicles (UGVs)					
Data analysis & evaluation technologies	Management zones delineation	All	All	Now available/available in the next 5 years (some technologies are already well established others are still under development).	Low to medium	Improving the Whole farm Information Management
	Decision support system					
	Farm management information system (FMIS)					
	Farm management, forecasting and crop monitoring software					
Guidance technology	Auto-guidance systems	All	All	Now available	Low	Operational Excellence / Improving the Whole-farm Information Management
	Control traffic farming					

Variable rate applications	Variable rate fertilizer application	All	All	Now available (automatic calibration)	Medium	Operational Excellence / License to operate / Improving the whole-farm information management.
				Available in the next 5 years (Online sensors for spread pattern determination)		
	Variable rate lime application			Now available	Low	Operational Excellence / Improving the Whole-farm Information Management.
	Variable rate manure application			Now available (flow control systems)	Low	Operational excellence / License to operate
				Available in the next 5 years (On-board nutrient sensing)		
	Variable rate pesticide application (Map-based system)			Now available	Medium	Operational Excellence/Improving the Whole farm information management
	Variable rate pesticide application (Real-time sensor based system)			Available now/available in the next 5 years (Some technologies are already well established others are still under development).	Medium to high	
	Boom height control			Now available		
	Variable rate planting/seeding		All (not in Orchards and Vineyards)		Medium	
	Precision physical weeding		All	Available in the next 5 years	N/A	Operational excellence / Improving the whole farm Information management
	Precision irrigation and irrigation scheduling			Now available	Low to medium	Operational excellence / Improving the whole farm Information management

Table 2.6 Smart farming technologies economic impact

SFT	Economic impact
Global navigation satellite systems (GNSS)	GNSS technologies do not have direct economic impact, but it is a requirement for most PA applications and therefore the technologies may have an indirect benefit.
Differential GNSS	
Real time kinematic (RTK)	
Network RTK (NRTK)	
Wide area RTK (WARTK)	
Undifferenced GNSS	
Precise point positioning (PPP)	
Fast PPP (FPPP)	
Elevation maps	Elevation, soil parameters (ECa, pH, moisture content) and yield mapping does not offer direct economic impact if it is not interpreted from Crop Consultants together with the farmer to apply site-specific crop management.
Soil mapping	
Yield mapping	
Yield monitor display	
RGB cameras	Imaging does not offer direct economic impact, but if it is used for VR application, the reduction in inputs will reflect on the farm income.
LiDAR sensors	
ToF (IR) cameras	
Light curtains	
Multi/Hyper-spectral Cameras	
Thermal cameras	
Spectral sensors	NDVI measurement does not offer direct economic impact, but if it is used for VR application, the reduction in inputs will reflect on the farm income.
Fluorescence sensors	
Frequency domain reflectometry	Soil moisture sensors measurements do not offer direct economic impact, but if it is used together with irrigation services they can be used for the reduction of irrigation water which reflects on the farm income.
Time domain reflectometry	
Amplitude domain reflectometry	
Phase transmission	
Time domain transmission	
Tensiometers	
Gipsum blocks	
Granular matrix sensors	
Heat dissipation sensors	

(continued)

Table 2.6 (continued)

SFT	Economic impact
Travel speed sensor	These technologies do not offer specific direct economic gains for the farmer, but they may have an indirect benefit when applied together with VR applications.
Tractor sensing systems using ISOBUS	
Unmanned aerial vehicles	Unmanned vehicles can provide profit for the farm, mainly due to limited/absent labour cost and less fuel costs compared with tractor mounted systems. Small low weight vehicles may reduce costs in relation to soil compaction and damage.
Unmanned ground vehicles	
Management zone delineation	Delineation of zones does not offer direct economic impact, except if they are taken into consideration for farm management optimization.
Decision support system	FMIS provides to farmers/farm managers detailed budgeting procedures, field planning, book-keeping for subsidy applications and for public authorities audits (Lawson et al. 2011). If data coming from SFTs of all kinds are imported in a general FMIS, then a series of documentation data will be able to automatically be developed, management time will be reduced significantly and due to improved management quality the farmer/farm manager will supply any regulatory body with detailed information from the farm that without the interference of SFTs would not be available (Steffe 2000).
Farm management information System (FMIS)	
Farm management, forecasting and crop monitoring software	
Auto-guidance systems	The economic benefit of using guidance technologies, either lightbar or auto-steer, comes from improved pass-to-pass efficiency and limitation of overlapping. This means that the applied agricultural inputs (seeds, fertilizers and pesticides) will be reduced with a direct positive impact on farm's economics. An example of such impact from application of guidance systems was given by Shockley et al. (2011) where the use of auto-steer improved performance during planting and fertilizer application leading to reduced inputs with cost benefit of approximately 2.4, 2.2 and 10.4% for seed, fertilizer and tractor fuel, respectively. Another case of auto-steer use was for peanut digging operations, where the traditional row deviation was 180 mm and when auto-steer was applied there were average net returns of 94–695 US$/ha depending on row deviations (Ortiz et al. 2013).
Controlled traffic farming (CTF)	CTF is based on the principle of using the same tramlines for all operations, which has a direct increase of farm's profit, because it reduce dramatically input costs related to farm machinery (time, fuel & machinery), while reducing significantly soil compaction with direct positive effect on crop yields. It has been seen that farms in Australia have cut their machinery costs by as much as 75% while their crop yields have risen. In the UK, the Colworth project showed that the fact that CTF reduce inputs results in healthier crops and soils showing also the sustainability profile of CTF.

(continued)

Table 2.6 (continued)

SFT	Economic impact
Variable rate fertilizer application	Crop yield is dependent on nitrogen supply. Sogaard and Kierkegaard (1994) showed that this relationship can be achieved using a quadratic equation, where the parabolic shape of this equation depicts that the marginal increase of yield decreases after a certain point with increased supply of nitrogen. The application of fertilizers reaches an economic optimum at some point depending on the site-specific soil type, precipitation etc. In principle, when a field is divided into management zones and fertilizers are applied near the economic optimum in these management zones, higher returns for the farm can be achieved. In addition, some studies have shown that when variable fertilizer application is used there is a reduction in the total amount of fertilizer used (Koch et al. 2004), reflecting in Nitrogen Use Efficiency (NUE) increase. Raun et al. (2001) found an average NUE increase of more than 15% in winter wheat in comparison to uniform application of fertilizer. More specifically, in-season application at 1 m^2 spatial resolution (based on optical sensing) increased their simple estimate of revenue (grain revenue minus fertilizer cost) by 11 US$/ha, when fertilizer was also applied before planting (fixed rate) and more than 28US$/ha when fertilizer was only applied in-season. Mamo et al. (2003) have seen that variable rate nitrogen application in corn increased farm profit by 8–23 US$/ha compared to uniform application. Although similar yields were found, less fertilizer was used. Other studies based on simulation models have shown moderate results from variable rate fertilizer application in cereals (Pedersen et al. 2003) and (Pedersen and Pedersen 2002). Variable rate fertilizer application requires both fixed and variable costs to be invested in, such as soil sampling, online sensing, delineation of management zones, VR equipment (e.g. GPS receiver, onboard computer, software, VR system, etc.). Farm size plays a significant role in absorbing these costs (economics of scale) and large farms can spread them over a larger area (Koch et al. 2004). Koch et al. (2004) found an increase of 25.6–38.6 US$/ha in net returns for VR nitrogen application on Colorado corn compared to uniform application rates, both in a farmer and custom applied scenario. Variable rate application based on grid soil sampling resulted in the lowest net return, primarily due to increased fertilizer uses and soil sampling costs. The highest returns for VR application are expected on fields with high and spatially variable nutrient requirements (Raun et al. 2001).

(continued)

Table 2.6 (continued)

SFT	Economic impact
Variable rate lime application	Variable rate application of lime, as opposed to uniform application, increases soil pH, reduces in-field variability and increases soybean yield (Weisz et al. 2003). While investigating VR lime application in four studies, 75% of the studies showed positive economic effect, while the rest 25% indicated mixed results. It have been seen that precise lime application has better results in legumes than in corn and wheat, because legumes respond to pH up to 6, while corn and wheat are limited to pH 5–5.5 (Weisz et al. 2003). The main cost of VR lime application is grid sampling. The actual amount of lime used depends on the soil variability, the sampling method and the sampling resolution, as well as on drought stress, environmental factors, variability level and acidity in the field, etc. (Weisz et al. 2003). VR lime application appears to be only profitable for high value crops (Swinton and Lowenberg-DeBoer 1998).
	Kuang et al. (2014) found an increase in lime consumption in Danish spring barley with simultaneous yield increase and net profit (US$4.1/ha) when VR approach was applied compared to the conventional approach. Weisz et al. (2003) performed grid sampling and VR lime application for 3 consecutive years in no-till soybean fields and found a net loss of US$12.99/ha compared to uniform lime application. However, when grid sampling was executed only in year 1 and 3, and performed the VR lime in each year (with year 2 based on the PH map of year 1) this turns into a net gain of US$4.86/ha over 3 years. Similarly, using the pH map from year 1 to apply lime for 3 years in the areas were lime was initially required leads to a net gain of US$7.31/ha. Like fertilizer, variable rate lime application requires investment in both fixed and variable costs, such as soil sampling, delineation of management zones, VR equipment (e.g. GPS receiver, on-board computer, software and VR system).
Variable rate manure application	Managing manure as fertilizer resource for crop production can increase the return for the producer and the overall production efficiency of an animal-crop farming system (Huber et al. 1993). Precision management of manure has the potential to further improve farming system production efficiency (Morris et al. 1999). The key to VR manure application in general is the existence of an application map, which is laborious and time consuming to generate when acquired without sensor technology (Schellberg and Reiner 2009). Although no literature is available considering the economic return of VR manure application, many similarities with VR granular (inorganic) fertilizer applications can be seen. The main difference is the fact that here the applied product is much more bulky, heterogeneous and lower in nutrient content and financial value. It should be noted that some VR manure systems can be retrofitted to the tanker that farmers already have (Brambilla et al. 2015).

(continued)

Table 2.6 (continued)

SFT	Economic impact
Variable rate pesticide application (Map-based system)	Site-specific pest management has costs as well as benefits. Costs of map-based VRA are attributed to mapping, data processing, decision making, and site-specific application technology. Commercial mapping services typically charge 4.5–9.0 €/ha to map field boundaries including waterways and other physical features (Batte and Ehsani 2006). Gerhards and Sökefeld (2003) estimated the costs of a direct injection system on 3.9 €/ha (in addition to the costs of the sprayer) for weed control in sugar beet, maize, winter wheat and winter barley in a German study. Batte and Ehsani (2006) stated that the extra cost of a precision controlled sprayer equipped with nozzles controlled individually based on GPS information would be about 8000 €. However, Timmermann et al. (2003) commented that several components of variable rate technology, including GPS, board computer and GIS, can also be used for other precision farming activities such as planting, fertilisation and harvest, being beneficial for other farming practices as well.
	Benefits in variable rate spraying are mainly associated with savings on pesticide use. Swinton (2003) stated that results on the likely profitability of site-specific weed management are uneven because certain studies focused on potential reduced cost from less herbicide spraying, while ignoring the increased capital cost of variable rate application equipment and the increased variable cost of information processing. Timmermann et al. (2003) found that the monetary savings resulting from the reduction in herbicide use varied between crops, depending on the amount of herbicides saved and the price of herbicide. In maize, winter wheat, winter barley and sugar beet, savings of respectively 42 €/ha, 32 €/ha, 27 €/ha, and 20 €/ha were realised. In this regard, savings also depend on the different economic thresholds for pest control and the different competitive power of the crops. Batte and Ehsani (2006) estimated spray material savings of about 4 €/ha for a map-based spraying system compared to a self-propelled sprayer without any form of GPS for guidance assistance or sprayer control. The magnitude of input savings further increased as waterways were added to the field. Those authors also calculated the costs of the spraying system. Most of the costs related to the fixed investment which diminished per hectare as farm size increased. They also concluded that the benefits will increase proportionally to the cost of the pesticide being applied and will increase with the number of annual applications and with the driver error-rate of the non-precision spraying system. Oriade et al. (1996) suggested that weed patchiness is the most important factor justifying the use of site-specific weed control. Using simulation, they show that economic and environmental benefits are almost zero at low weed pressures, particularly if weeds are evenly spread. The benefits were larger as weed populations and level of patchiness increased. At high weed patchiness, return values of 17 €/ha to 33 €/ha were found in corn and soybean. The authors concluded that returns from site-specific management less than 14 €/ha are not sufficient to warrant the practice. The costs of information collection, time effects, and human capital were not considered in this model.

(continued)

Table 2.6 (continued)

SFT	Economic impact
Variable rate pesticide application (Real-time sensor based system)	As with map-based VRA, benefits are mainly associated with savings on pesticides (especially herbicides and to some extend fungicides) use and these savings depend on crop type, pest distribution, pesticide price, etc. In contrast to map-based VRA, an additional step of generating an application map with the help of GIS is not necessary. Therefore, there are no additional costs for computers, GIS software or DGPS. However, the sensor technology can be very expensive, although cheap sensors are available as well. Gerhards and Sökefeld (2003) estimated the cost of a camera system for weed detection to 40,000 €, whereas Dammer and Wartenberg (2007) used an optoelectronic weed sensor of about 2000 €. The latter could however not distinguish between crops and weeds and was therefore limited to operations within the tramlines. Gerhards and Sökefeld (2003) evaluated the economic benefits of a real-time, automatic, site-specific weed control system compared to conventional field spraying. They found that although the costs for the VRA technology were larger (9.56 €/ha vs. 5.20 €/ha), the average costs for weed control were lower due to herbicide savings (32 €/ha vs. 68 €/ha in winter wheat and winter barley, 69 €/ha vs. 148 €/ha in sugar beet, and 96 €/ha vs. 103 €/ha in maize). Based on these economic calculations, Dammer and Wartenberg (2007) comment that if sensors were available on the market, it would be profitable for farmers to invest in variable rate technologies. In summary, the net benefit of variable rate pesticide application will depend on the crop value, weed distribution, cost of pest/weed mapping, sensor systems as well as pesticide prices etc.
Boom height control	The uneven distribution due to boom height differentiation in the field may results in yield losses or in additional pesticide costs, however, no studies were found that calculated these economic effects.

(continued)

Table 2.6 (continued)

SFT	Economic impact
Variable rate planting/seeding	Soil variability is the main factor driving the economic performance of variable-rate seeding/planting. The return on investment of VR planting/seeding is low in very uniform fields, while in heterogeneous fields with clear high and especially low crop performance zones, the return on investment will be much higher. In the early years of VR planting/seeding development, its economic impact was unclear. In 1998, Bullock et al. (1998) observed differences in economically optimal plant densities as a function of yield potential in an extensive study in the Corn Belt region of the US, but they concluded that variable rate seeding would be infeasible, because of the high cost associated with characterizing site variability. Lowenberg-DeBoer (1998) stated that the investments necessary for adopting variable rate corn seeding would only be economically justifiable for farmers with some low yield potential land, but not for farmers with a mix of solely medium and high potential land. Taylor and Staggenborg (2000) concluded that variable rate seeding was only economically feasible on their fields of study if less expensive ways to generate the prescription map was available or if corn showed a greater yield response to seeding rate. In 2004, Shanahan et al. (2004) stated that "site-specific management of plant densities may be feasible", most likely due to technological advances. Dillon et al. (2009) performed sensitivity analysis with respect to alternative soils, seed price, wheat price and cost of VR seeding technology to determine its economic feasibility and concluded that the practice of VR seeding of wheat in France is feasible.
Precision physical weeding	As this technology is still in its infancy, no specific economic impact figures are readily available. However, a significant reduction of manual labour during physical weeding can be expected, especially in organic agriculture, which may lead to significant cost reductions.
Precision irrigation and irrigation scheduling	Lambert and Lowenberg-DeBoer (Lambert and Lowenberg-Deboer 2000) reported a possible economic benefit in corn yield and water use when using VR irrigation, but it was not described in numbers. Many authors (Booker et al. 2015; Colaizzi et al. 2009; Evans and King 2012) have mentioned the high costs of such systems, and a beneficiary in yield, work load, water use, pesticide use, etc. especially in climatic unfavourable years. However, numbers or comparisons are not given.

www.controlledtrafficfarming.com

References

Abidine AZ, Heidman BC, Upadhyaya SK, Hills DJ (2002) Application of RTK GPS based autoguidance system in agricultural production. Paper No. O21152. ASAE, St. Joseph, MI, USA

Adamchuk VI, Viscarra Rossel RA (2014) Precision agriculture: proximal soil sensing. Encyclopedia of Agrophysics

Adamchuk VI, Hummel JW, Morgan MT, Upadhyaya SK (2004) On-the-go soil sensors for precision agriculture. Comput Electron Agric 44:71–91

AGROBOT (2012) AGROBOT – strawberry harvesters. http://www.agrobot.com

Akdemir B, Sağlam N, Bellitürk K, Aydogdu B (2007) Development a variable rate controller for centrifugal fertiliser (Design the system), 4th International Symposium on Intelligent information technology in Agriculture, 26–29 October 2007, Beijing, China

Andrade-Sanchez P, Gore MA, Heun JT, Thorp KR, Carmo-Silva AE, French AN, Salvucci ME, White JW (2014) Development and evaluation of a field-based high-throughput phenotyping platform. Funct Plant Biol 41:68–79

Arnó J, Martinez-Casasnovas J, Ribes-Dasi M, Rosell JR (2009) Review: precision viticulture. research topics, challenges and opportunities in site-specific vineyard management. Span J Agric Res 7:779–790

Arnó J, Escolà A, Valle JM, Llorens J, Sanz R, Masip J, Palacın J, Rosell-Polo JR (2013) Leaf area index estimation in vineyards using a ground-based LiDAR scanner. Precis Agric 14:290–306

Arnó J, Escolà A, Masip J, Rosell-Polo JR (2015) Influence of the scanned side of the row in terrestrial laser sensor applications in vineyards: practical consequences. Precis Agric 16(2):119–128

Åstrand B, Baerveldt AJ (2002) An agricultural mobile robot with vision-based perception for mechanical weed control. Auton Robot 13(1):21–35

Backman J, Oksanen T, Visala A (2013) Applicability of the ISO 11783 network in a distributed combined guidance system for agricultural machines Original Research Article. Biosyst Eng 114(3):306–317

Baio FHR, Moratelli RF (2011) Auto guidance accuracy evaluation and contrast of the operational field capacity on the mechanized plantation system of sugar cane. Eng Agric 31:367–375

Batte MT, Ehsani MR (2006) The economics of precision guidance with auto-boom control for farmer-owned agricultural sprayers. Comput Electron Agric 53(1):28–44

Behic Tekin A, Okyay Sındır K (2013) Variable rate control system designed for spinner disc fertilizer spreader–"Pre Fer". Agric Eng 2:45–53

Benavente E, García-Toledano L, Carrillo JM, Quemada M (2013) Thermographic imaging: assessment of drought and heat tolerance in Spanish germplasm of Brachypodium distachyon. Procedia Environ Sci 19:262–266

Berne D (2015) Agricultural irrigation initiative: overview of center pivot irrigation systems. https://neea.org/docs/default-source/reports/overview-of-center-pivot-irrigation-systems.pdf?sfvrsn=4. Retrieved on 10 Jan 2016

Berni JA, Zarco-Tejada PJ, Suarez L, Fereres E (2009) Thermal and narrowband multispectral remote sensing for vegetation monitoring from an unmanned aerial vehicle. IEEE Trans Geosci Remote Sens 47:722–738

Blackmore BS (2007) A systems view of agricultural robots. In: Proceedings of the 6th European conference on precision agriculture, pp 23–31

Booker JD, Lascano RJ, Molling CC, Zartman RE, Acosta-Martinez (2015) Temporal and spatial simulation of production-scale irrigated cotton systems. Precis Agric 16:630–653

Brambilla M, Calcante A, Oberti R, Bisaglia C (2015) Slurry tanker retrofitting with variable rate dosing system: a case study. In: Precision agriculture '15. Wageningen Academic Publishers, pp 123–135

Bramley RGV (2001) Progress in the development of precision viticulture-variation in yield, quality and soil properties in contrasting Australian vineyards. In: Currie LD, Loganathan P (eds) Precision tools for improving land management, Occasional report No. 14. Fertilizer and Lime Research Centre/Massey University, Palmerston North, pp 25–43

Bullock DG, Bullock DS, Nafziger ED, Doerge TA, Paszkiewicz SR, Carter PR, Peterson TA (1998) Does variable rate seeding of corn pay? Agron J 90:830–836
Calcante A, Brambilla M, Oberti R, Bisaglia C (2015) A retrofit variable-rate control system for pressurized slurry tankers. Appl Eng Agric 31(4):569–579
Calderón R, Navas-Cortés JA, Lucena C, Zarco-Tejada PJ (2013) High-resolution airborne hyperspectral and thermal imagery for early detection of Verticillium wilt of olive using fluorescence, temperature and narrow-band spectral indices. Remote Sens Environ 139:231–245
Camp CR (1998) Subsurface drip irrigation: a review. Applied engineering in agriculture. American Society of Agricultural and Biological Engineers 41(5):1353–1367
Cerovic ZG, Samson G, Morales F, Tremblay N, Moya I (1999) Ultraviolet-induced fluorescence for plant monitoring: present state and prospects. Agronomie 19:543–578
Chen S, Shiping Z (2011) Experimental research on variable rate fertization technology of spinning disc spreader for rice. In: International conference on agricultural and Biosystems Engineering
Chene Y, Rousseau D, Lucidarme P, Bertheloot J, Caffier V, Morel P, Belin E, Chapeau-Blondeau F (2012) On the use of depth camera for 3D phenotyping of entire plants. Comput Electron Agric 82:122–127
Christen D, Schönmann S, Jermini M, Strasser RJ, Défago G (2007) Characterisation and early detection of grapevine stress responses to esca disease by in situ chlorophyll fluorescence and comparison with drought stress. Environ Exp Bot 60:504–514
Colaizzi PD, Gowda PH, Marek TH, Porter O (2009) Irrigation in the Texas high plains: a brief history and potential reductions in demand. Irrig Drain 58:257–274
Dammer KH, Wartenberg G (2007) Sensor-based weed detection and application of variable herbicide rates in real time. Crop Prot 26(3):270–277
Darr M (2012) CAN Bus technology enables advanced machinery management. Resour Eng Technol Sustainable World 19(5):10–11
Dillon CR, Gandonou J, Shockley J (2009) Variable rate seeding for French wheat production: profitability and production risk management potential. In: Lokhorst C, Huijsmans JFM, de Louw RPM (eds) JIAC2009 book of abstracts. Wageningen Academic Publishers, Wageningen, p 350
Doruchowski G, Swiechowski W, Holownicki R, Godyn A (2009) Environmentally-dependent application system (EDAS) for safer spray application in fruit growing. J Hortic Sci Biotechnol (special issue): 107–112
Ehlert D, Heisig M, Adamek R (2010) Suitability of a laser rangefinder to characterize winter wheat. Precis Agric 11(6):650–663
Erdle K, Mistele B, Schmidhalter U (2011) Comparison of active and passive spectral sensors in discriminating biomass parameters and nitrogen status in wheat cultivars. Field Crop Res 124:74–84
Ess DR, Parsons SD, Medlin CR (2001) Implementing site-specific management: sprayer technology – controlling application rate on the go. http://www.ces.purdue.edu/extmedia/AE/SSM-5-W.pdf
Evans RG, King BA (2012) Site-Specific sprinkler irrigation in a water-limited future. Applied engineering in agriculture. Am Soc Agric Biol Eng 55(2):493–504
Fanourakis D, Briese C, Max JFJ, Kleinen S, Putz A, Fiorani F, Ulbrich A, Schurr U (2014) Rapid determination of leaf area and plant height by using light curtain arrays in four species with contrasting shoot architecture. Plant Methods 10:9
Feng W, Yao X, Zhu Y, Tian YC, Cao WX (2008) Monitoring leaf nitrogen status with hyperspectral reflectance in wheat. Eur J Agron 28:394–404
Forouzanmehr E, Loghavi M (2012) Design, development and field evaluation of a map-based variable rate granular fertilizer application control system. Agric Eng Int CIGR J 14(4):255–261
Foth HD, Ellis BG (1988) Soil fertility. Wiley, New York
Fountas S, Blackmore S, Ess D, Hawkins S, Blumhoff G, Lowenberg-Deboer J, Sorensen CG (2005) Farmer experience with precision agriculture in Denmark and the US Eastern corn belt. Precis Agric 6:121–141

Fountas S, Wulfsohn D, Blackmore S, Jacobsen HL, Pedersen SM (2006) A model of decision making and information flows for information-intensive agriculture. Agric Syst 87:192–210
Fountas S, Balafoutis A, Anastasiou E, Koundouras S, Kotseridis G, Kallithraka E, Kyraleou M (2014) Site-specific variability of grape composition and wine quality. Proceedings of International Conference on Precision Agriculture, Sacramento
Fountas S, Carli C, Sørensen CG, Tsiropoulos Z, Cavalaris C, Vatsanidou A, Liakos B, Canavari M, Wiebensohn J, Tisserye B (2015) Farm management information systems: current situation and future perspectives. Comput Electron Agric 115:40–50
Fulton JP, Shearer SA, Chabra G, Higgins SF (2001) Performance assessment and model development of a variable-rate, spinner-disc fertilizer applicator. Trans Am Soc Agric Eng 44(5):1071–1082
Funk TL, Robert MJ (2003) Variable rate manure spreader: technology to validate a nutrient management plan. http://livestocktrail.illinois.edu/uploads/sowm/papers/p111-121.pdf
Gan-Mor S, Clark RL, Upchurch BL (2007) Implement lateral position accuracy under RTK-GPS tractor guidance. Comput Electron Agric 59(1):31–38
Gerhards R, Sökefeld M (2003) Precision farming in weed control – sytem components and economic benefits. In: Stafford J, Werner A (eds) Precision agriculture. Wageningen Academic Publishers, Wageningen, pp 229–234
Ghozlen NB, Cerovic ZG, Germain C, Toutain S, Latouche G (2010) Non-destructive optical monitoring of grape maturation by proximal sensing. Sensors 10:10040–10068
Gravalos I, Tsiropoulos Z, Kateris D, Gialamas T, Xyradakis P, Avgoustis A, Fountas S (2014) Soil moisture remote monitoring from an agricultural tractor. In: Second international conference on robotics and associated high-technologies and equipment for agriculture and forestry. Madrid, 22–23
Grevis-James IW, DeVoe DR, Bloome PD, Batchelder DG, Lambert BW (1983) Microcomputer-based data acquisition for tractors. Trans ASABE 26(3):692–695
Grisso R, Alley M, Thomason W, Holshouser D, Roberson GT (2011) Precision farming tools: variable-rate application. Va Coop Ext Publ:442–505
Harter DD, Kaufman KR (1979) Microprocessor based data acquisition system for tractor tillage measurements. Paper No. 79–5026, ASAE, St Joseph, MI, USA
HAU & CERETETH (2013) Dionysus Robot to Help Winegrowers. http://vimeo.com/68527042
Hernández-Pajares M, Juan JM, Sanz J (2000) Improving the Abel inversion by adding ground GPS data to LEO radio occultations in ionospheric sounding. Geophys Res Lett 27(16):2473–2476
Héroux P, Kouba J (1995) GPS precise point positioning with a difference. Natural Resources Canada, Geomatics Canada, Geodetic Survey Division
Herwitz SR, Johnson LF, Dunagan SE, Higgins RG, Sullivan DV, Zheng J, Lobits BM, Leung JG, Gallmeyer BA, Aoyagi M, Slye RE, Brass JA (2004) Imaging from an unmanned aerial vehicle: agricultural surveillance and decision support. Comput Electron Agric 44:49–61
Hoffmeister D, Waldhoff G, Korres W, Curdt C, Bareth G (2016) Crop height variability detection in a single field by multi-temporal terrestrial laser scanning. Precis Agric 17(3):296–312
Hosoi F, Omasa K (2012) Estimation of vertical plant area density profiles in a rice canopy at different growth stages by high-resolution portable scanning lidar with a lightweight mirror. ISPRS J Photogram Remote Sens 74:11–19
Hostens I, Anthonis J, Kennes P, Ramon H (2000) Six-degrees-of-freedom test rig design for simulation of mobile agricultural machinery vibrations. J Agric Eng Res 77(2):155–169
Huber DM, Sutton AL, Jones DD, Joern BC, Mitchell JK (1993) Nutrient management of manure to enhance crop production and protect the environment. In: Proceedings of the Integrated Resource Management and Landscape Modifications for Environmental Protection Conference, Chicago, IL. American Society of Agricultural Engineers, St. Joseph, MI
Iftikhar N, Pedersen TB (2011) Flexible exchange of farming device data. Comput Electron Agric 75(1):52–63

Inman D, Khosla R, Mayfield T (2005) On-the-go active remote sensing for efficient crop nitrogen management. Sens Rev 25(3):209–214

Jorgensen RN, Sorensen CG, Maagaard J, Havn I, Jensen K, Sogaard HT, Sorensen LB (2007) Hortibot: a system design of a robotic tool carrier for high-tech plant nursing. CIGR J IX

Karkee M, Steward B, Kruckeberg J (2013) Automation of pesticide application systems. In: Zhang G, Pierce FJ (eds) Agricultural automation: fundamentals and practices. CRC Press, Boca Raton

Kelly I, Melhuish C (2001) Slugbot: a robot predator. In: Advances in artificial life. Springer, Berlin/Heidelberg, pp 519–528

Kitchen NR, Sudduth KA, Myers DB, Drummond ST, Hong SY (2005) Delineating productivity zones on claypan soil fields apparent soil electrical conductivity. Comput Electron Agric 46:285–308

Koch B, Khosla R, Frasier WM, Westfall DG, Inman D (2004) Economic feasibility of variable-rate nitrogen application utilizing site-specific management zones. Agron J 96(6):1572–1580

Kuang B, Tekin Y, Waine T, Mouazen AM (2014) Variable rate lime application based on on-line visible and near infrared (vis-NIR) spectroscopy measurement of soil properties in a Danish field. Ageng 2014

Lambert D, Lowenberg-De Boer J (2000) Precision agriculture profitability review. Purdue University

Landau H, Chen X, Kipka A, Vollath U (2007) Latest developments in Network RTK modeling to support GNSS modernization. Positioning 1(11)

Lawson LG, Pedersen SM, Sorensen CG, Pesonen L, Fountas S, Werner A, Oudshoorn FW, Herold L, Chatzinikos T, Kirketerp IM, Blackmore S (2011) A four nation survey of farm information management and advanced farming systems: a descriptive analysis of survey responses. Comput Electron Agric 77:7–20

Li X, Zhang X, Ren X, Fritsche M, Wickert J, Schuh H (2015) Precise positioning with current multi-constellation global navigation satellite systems: GPS, GLONASS, Galileo and BeiDou. Sci Rep 5:8328

Liu C, Liu W, Lu X, Ma F, Chen W, Yang J, Zheng L (2014) Application of multispectral imaging to determine quality attributes and ripeness stage in strawberry fruit. PLoS One 9(2):e87818

Lowenberg-DeBoer JM (1998) Economics of variable rate planting for corn. In: Robert PC, Rust RH, Larson WE (eds) Proceedings of 4th international conference on precision agriculture. Purdue University, Saint Paul, pp 1643–1651

Maleki MR, Mouazen AM, De Ketelaere B, Ramon H, De Baerdemaeker J (2008a) On-the-go variable-rate phosphorus fertilisation based on a visible and near-infrared soil sensor. Biosyst Eng 99(1):35–46

Maleki MR, Ramon H, De Baerdemaeker J, Mouazen AM (2008b) A study on the time response of a soil sensor-based variable rate granular fertiliser applicator. Biosyst Eng 100(2):160–166

Mamo M, Malzer GL, Mulla DJ, Huggins DR, Strock J (2003) Spatial and temporal variation in economically optimum nitrogen rate for corn. Agron J 95(4):958–964

Martelloni L (2014) Design and realization of an innovative automatic machine able to perform site-specific thermal weed control in maize. PhD thesis, Università degli Studi di Firenze

Martínez JL, Mandow A, Morales J, Pedraza S, García-Cerezo A (2005) Approximating kinematics for tracked mobile robots. Int J Robot Res 24(10):867–878

McBratney AB, Whelan B, Ancev T, Bouma J (2005) Future directions of precision agriculture. Precis Agric 6:7–23

Montes JM, Techow F, Dhillon BS, Mauch F, Melchinger AE (2011) High-throuput non-destructive biomass determination during early plant development in maize under field conditions. Field Crop Res 121:268–273

Morris DK, Ess DR, Hawkins SE, Parsons SD (1999) Development of a site-specific application system for liquid animal manures. Appl Eng Agric 15(6):633–638

Munoz-Carpena R (2017) Field devices for monitoring soil water content. http://edis.ifas.ufl.edu/pdffiles/AE/AE26600.pdf. Retrieved 20 Feb 2017

Nakarmi AD, Tang L (2012) Automatic inter-plant spacing sensing at early growth stages using a 3D vision sensor. Comput Electron Agric 82:23–31
Nikkila R, Seilonen I, Koskinenet K (2010) Software architecture for farm management information systems in precision agriculture. Comput Electron Agric 70(2):328–336
Oriade CA, King RP, Forcella F, Gunsolus JL (1996) A bioeconomic analysis of site-specific management for weed control. Rev Agric Econ 18:523–535
Ortiz BV, Balkcom KB, Duzy L, van Santen E, Hartzog DL (2013) Evaluation of agronomic and economic benefits of using RTK-GPS-based auto-steer guidance systems for peanut digging operations. Precis Agric 14:357–375
Paulus S, Schumann H, Kuhlmann H, Leon J (2014) High-precision laser scanning system for capturing 3D plant architecture and analyzing growth of cereal plants. Biosyst Eng 121:1–11
Pedersen SM (2003) Precision farming, technology assessment of site-specific input application in cereals. PhD dissertation, Technical University of Denmark, p 343
Pedersen SM, Pedersen JL (2002) Economic and environmental impact of site-specific N-application – based on different weather conditions and arable crops. Paper presentation at the 6th international conference on precision agriculture, Jul 14–17, 2002, Minneapolis, US
Pedersen SM, Fountas S, Blackmore BS, Gylling M, Pedersen JL, Pedersen HH (2003) Adoption of precision farming in Denmark. 4th European conference on precision agriculture proceedings, p 533–538
Perez-Ruiz M, Slaughter DC, Gliever C, Upadhyaya SK (2013) Tractor-based Real-time Kinematic-Global Positioning System (RTK-GPS) guidance system for geospatial mapping of row crop transplant. Biosyst Eng 111:64–71
Ramirez DA, Yactayo W, Gutierrez R, Mares V, De Mendiburu F, Posadas A, Quiroz R (2014) Chlorophyll concentration in leaves is an indicator of potato tuber yield in water-shortage conditions. Sci Hortic 168:202–209
Raun W, Solie J, Johnson G, Stone M, Mullen R, Freeman K, Thomason W, Lukina E (2001) Improving Nitrogen use efficiency in cereal grain production with optical sensing and variable rate application. Agron J 94:815–820
Remondi BW (1985) Performing centimeter-level surveys in seconds with GPS carrier phase: initial results. Navigation 32(4):386–400
Reutebuch SE, McGaughey RJ, Andersen HE, Carson WW (2003) Accuracy of a high-resolution lidar terrain model under a conifer forest canopy. Can J Remote Sens 29(5):527–535
Rojo B, Fabio H (2012) Evaluation of an auto-guidance system operating on a sugar cane harvester. Precis Agric 13:141–147
Ramon H, Missotten B, De Baerdemaeker J (1997) Spray boom motions and spray distribution: Part 2, Experimental validation of the mathematical relation and simulation results. J Agric Eng Res 66(1):31–39
Rosell JR, Llorens J, Sanz R, Arno J, Ribes-Dasi M, Masip J, Escola A, Camp F, Solanelles F, Gracia F, Gil E, Val L, Planas S, Palacin J (2009) Obtaining the three-dimensional structure of tree orchards from remote 2D terrestrial LIDAR scanning. Agric For Meteorol 149:1505–1515
Rouse JW, Haas RH, Schell JA, Deering DW (1973) Monitoring vegetation systems in the Great Plains with ERTS, Third ERTS symposium, NASA SP351 I, pp 309–317
Ruckelshausen A, Biber P, Dorna M, Gremmes H, Klose R, Linz A, Rahe R, Resch R, Thiel M, Trautz D, Weiss U (2009) BoniRob: an autonomous field robot platform for individual plant phenotyping. Precis Agric 9:841
Scarlett AJ (2001) Integrated control of agricultural tractors and implements: a review of potential opportunities relating to cultivation and crop establishment machinery. Comput Electron Agric 30:167–191
Schellberg J, Reiner L (2009) A site-specific slurry application technique on grassland and on arable crops. Bioresour Technol 100:280–286
Shanahan JF, Doerge TA, Johnson JJ, Vigil MF (2004) Feasibility of site-specific management of corn hybrids and plant densities in the great plains. Precis Agric 5:207–225

Shinners TJ, Digman MF, Panuska JC (2012) Overlap loss of manually and automatically guided mowers. Appl Eng Agric 28:5–8
Shockley JM, Dillon CR, Stombaugh T (2011) A whole farm analysis of the influence of auto-steer navigation on net returns, risk and production practices. J Agric Appl Econ 43(1):57–75
Society of Automotive Engineers (SAE) (1995) Surface vehicle recommended practice J 1939-7x
Sogaard HT, Kierkegaard P (1994) Yield reduction resulting from uneven fertiliser distribution. Trans ASAE 37:1749–1752
Sørensen CG (1999) A Bayesian network based decision support system for the management of field operations. Case: harvesting operations. Ph.D. thesis, Technical University of Denmark, 193 pp
Sørensen GC, Fountas S, Nash E, Pesonen L, Bochtis D, Pedersen SM, Basso B, Blackmore SB (2010) Conceptual model of a future farm management information system. Comput Electron Agric 72:37–47
Stafford JV (2000) Implementing precision agriculture in the 21st century. J Agric Eng Res 76(3):267–275
Steffe J (2000) Evolution of the farm environment: the need to produce a general information system. In: Beers G, Poppe KJ, de Putter I (eds) Agenda 2000 and the FADN agenda, Workshop report, Project code 63403. Agricultural Economics Research Institute (LEI), The Hague, pp 88–97
Swinton SM (2003) Site-specific pest management. In: den Hond F, Groenewegen P, van Straalen NM (eds) Pesticides – problems, improvements, alternatives. Blackwell Science, Oxford. 155 p
Swinton SM, Lowenberg-DeBoer J (1998) Evaluating the profitability of site-specific farming. J Prod Agric 11(4):439–446
Tabile RA, Godoy EP, Pereira RR, Tangerino GT, Porto AJ, Inamasu RY (2011) Design and development of the architecture of an agricultural mobile robot. Engenharia Agricola 31(1):130–142
Tagarakis A, Liakos V, Fountas S, Koundouras S, Aggelopoulou K, Gemtos T (2011) Management zones delineation using fuzzy clustering techniques in vines. In: Stafford J (ed) Precision agriculture, proceedings of the 8th European conference on precision agriculture. Wageningen Academic Publishers, Wageningen, pp 191–200
Taylor RK, Staggenborg S (2000) Using a GIS to evaluate the potential of variable rate corn seeding. ASAE meeting presentation
Taylor RK, Schrock MD, Staggenborg SA (2002) Extracting machinery management information from GPS data. Paper No. 02–10008. St. Joseph, Michigan: ASABE
Tekin AB, Yurdem H, Fountas S, Tsiropoulos Z, Aygun T (2013) Design and implementation of robotic platform: RoboTurk. In: Proceedings of the EFITA WCCA-CIGR conference "Sustainable Agriculture through ICT Innovation"
Thoren D, Thoren P, Schmidhalter U (2010) Influence of ambient light and temperature on laser-induced chlorophyll fluorescence measurements. Eur J Agron 32(2):169–176
Thorp KR, Dierig DA (2011) Color image segmentation approach to monitor flowering in lesquerella. Ind Crop Prod 34:1150–1159
Timmermann C, Gerhards R, Kühbauch W (2003) The economic impact of site-specific weed control. Precis Agric 4:249–260
Tressos IAK (2011) Zeus: codename savage. http://www.savage.gr/gallery.html
Tsiropoulos Z, Fountas S, Gemtos T, Gravalos I, Paraforos D (2013) Management information system for spatial analysis of tractor-implement draft forces. In: Precision agriculture'13. Wageningen Academic Publishers, Wageningen, pp 349–356
Tsiropoulos Z, Fountas S, Gravalos I, Augoustis A, Arslan S, Misiewicz P, Gemtos T (2015) Importance of measuring tillage implement forces for reduced fuel consumption and increased efficiency without affecting tillage depth. In: Precision agriculture'15. Wageningen Academic Publishers, Wageningen, pp 217–232
Tu X (2013) Robust navigation control and headland turning optimization of agricultural vehicles. Graduate theses and dissertations, Paper 13188. Iowa State University

Van Elderen E (1977) Heuristic strategy for scheduling farm operations. Centre for Agricultural Publishing and Documentation, Wageningen, .The Netherlands, 217 pp

Van Liedekerke PJ, Baerdemaeker D, Ramon H (2006) Precision Agriculture: Fertilizer application control. In: Munack A (ed) CIGR handbook of agricultural engineering VOLUME VI information technology. CIGR, St. Joseph

Verdu S, Ivorra E, Sanchez AJ, Giron J, Barat JM, Grau R (2013) Comparison of TOF and SL techniques for in-line measurement of food item volume using animal and vegetable tissues. Food Control 33:221–226

Vlugt P (2013) Ag Industry's initiative in electronic standards implementation. http://www.clubofbologna.org/ew/documents/1.3_Van_der_Vlugt.pdf

Vollmann J, Walter H, Sato T, Schweiger P (2011) Digital image analysis and chlorophyll metering for phenotyping the effects of nodulation in soybean. Comput Electron Agric 75:190–195

Wang W, Li C (2014) Size estimation of sweet onions using consumer-grade RGB-depth sensor. J Food Eng 142:153–162

Weisz R, Heiniger R, White JG, Knox B, Reed L (2003) Long-term variable rate lime and phosphorus application for Piedmont no-till field crops. Precis Agric 4(3):311–330

Whelan B, Taylor J (2013) Precision agriculture for grain production systems. CSIRO Publishing, Collingwood

Wollenhaupt NC, Mulla DJ, Gotway Crawford CA (1997) Chapter 2: Soil sampling and interpolation techniques for mapping spatial variability of soil properties. In: Pierce FT, Sadler EJ (eds) The state of site-specific management for agriculture. ASA-CSSA-SSSA, Madison, pp 19–53

Yahya A, Zohadiea M, Kheiralla AF, Giewa SK, Boona NE (2009) Mapping system for tractor-implement performance. Comput Electron Agric 69:2–11

Yigit CO (2016) Experimental assessment of post-processed kinematic precise point positioning method for structural health monitoring. Geomat Nat Haz Risk 7(1):360–383

Zarco-Tejada PJ, Gonzales-Dugo V, Williams LE, Suarez L, Berni JAJ, Goldhamer D, Fereres E (2013a) A PRI-based water stress index combining structural and chlorophyll effects: assessment using diurnal narrow-band airborne imagery and the CSWI thermal index. Remote Sens Environ 138:38–50

Zarco-Tejada PJ, Guillen-Climent MI, Hernandez-Clemente R, Catalina A, Gonzalez MR, Martin P (2013b) Estimating leaf carotenoid content in vineyards using high resolution hyperspectral imagery acquired from an unmanned aerial vehicle (UAV). Agric For Meteorol 171–172:281–294

Zarco-Tejada P Hubbard N, Loudjani P (2014) Precision agriculture: an opportunity for EU farmers – potential support with the CAP 2014–2020. Joint Research Centre (JRC) of the European Commission. Monitoring Agriculture ResourceS (MARS) Unit H04, Brussels, Belgium

Zhang C, Kovacs JM (2012) The small unmanned aerial systems for precision agriculture: a review. Precis Agric 13:693–712

Zumberge JF, Heflin MB, Jefferson DC, Watkins MM, Webb FH (1997) Precise point positioning for the efficient and robust analysis of GPS data from large networks. J Geophys Res Solid Earth 102(B3):5005–5017

Chapter 3
Economic Potential of Site-Specific Fertiliser Application and Harvest Management

Markus Gandorfer and Andreas Meyer-Aurich

Abstract Site-specific fertiliser management has been discussed as an information-based farming concept that uses plant- and soil-specific information. However, agricultural practice has shown that, because of limited profitability, the adoption of site-specific fertiliser management often does not meet expectations. This chapter describes a framework for the economic assessment of site-specific fertiliser application and harvest management, provides an overview of selected studies and shows the future perspective of the technologies.

We concluded that precision farming technologies that aim to identify the economically optimal input rate (e.g. site-specific fertiliser application) often fail to provide considerable economic advantages for the farmer. This phenomenon can be explained by flat payoff functions, which are relevant for many agricultural production processes. Economically more promising from a theoretical point of view are precision farming approaches that enable higher product prices by achieving specific product qualities (e.g. site-specific harvest management). However, available studies currently do not provide empirical support for this theoretical conclusion.

Keywords Economic evaluation • Nitrogen fertiliser • Precision farming • Smart farming • Site-specific production functions

M. Gandorfer (✉)
Chair of Economics of Horticulture and Landscaping, Technical University of Munich, Freising, Germany
e-mail: markus.gandorfer@tum.de

A. Meyer-Aurich
Leibniz-Institute for Agricultural Engineering and Bioeconomy e.V, Potsdam, Germany
e-mail: ameyer@atb-potsdam.de

S.M. Pedersen, K.M. Lind (eds.), *Precision Agriculture: Technology and Economic Perspectives*, Progress in Precision Agriculture,
https://doi.org/10.1007/978-3-319-68715-5_3

3.1 Introduction

During the last two decades, the emergence of information technologies and geographic information systems has triggered technological solutions for information-based agricultural management systems. These systems have the potential to increase the efficiency of agricultural systems and contribute to economic and environmental gains. Because of the economic and environmental potential of these systems, two recent reports analyse precision farming technologies from a policy perspective. Because of a request from the European Parliament's Committee on Agriculture and Rural Development, Zarco-Tejada et al. (2014) published a study titled: "Precision Agriculture: An opportunity for EU farmers – Potential support with the CAP 2014–2020". The Organisation for Economic Co-operation and Development (OECD) has released a report discussing "Farm Management Practices to Foster Green Growth" where the question was raised: "Is precision agriculture the start of a new revolution?" (OECD 2016, p. 137). Despite the great interest in precision farming technologies shown by politicians and by researchers, the adoption of many of these technologies is still limited (OECD 2016). Some of the major constraints for adoption of precision farming technologies are the complexity of the technology (including incompatibility of components), time requirements, high start-up costs, and lack or uncertainty of profitability, among others (Griffin et al. 2004; Khanna et al. 1999; Reichardt et al. 2009; Robertson et al. 2012). Therefore, the aim of this chapter is to analyse the economics of different precision farming technologies as a major determinant for their adoption. Studies addressing different aspects of the economic potential of site-specific fertiliser application and harvest management were analysed, including the potential for using information technologies in farm management. The remaining part of this chapter is organised as follows. In Sect. 3.2, a framework for the economic assessment of site-specific fertiliser application and harvest management is presented. In the section that follows, selected studies that analyse the economic benefits of site-specific fertiliser application and harvest management are discussed. The chapter closes with some general conclusions.

3.2 Framework for Economic Assessment

According to Meyer-Aurich et al. (2008) a comprehensive economic assessment of precision farming technologies at the farm level needs to take into account all relevant monetary and non-monetary aspects, including effects on crop yield, input use, changes in management and the quality of work. Table 3.1 provides an overview of the costs and possible benefits of precision farming technologies. Four different types of costs that arise from farm-level implementation of precision farming technologies can be distinguished. Information costs are associated with the necessary investment in technologies or equipment rental fees necessary to ascertain specific information.

Table 3.1 Costs and possible benefits from Precision Farming

Costs of precision farming	Possible benefits
Information costs	Crop yield effects
Costs of data processing	Changes in input use
Costs of adapted management	Changes in management
Learning costs	Quality of work
	Production risk mitigation

Source: Meyer-Aurich et al. (2008), modified

Costs for data processing include costs for specific software or hardware products, but also opportunity cost for time needed to develop site-specific management schemes. Precision farming involves a change in management which may incur specific costs. In addition, learning costs, including opportunity costs because of inefficient use of precision farming technologies, need to be taken into account, particularly in the introduction phase.

Possible benefits of precision farming stem from crop yield effects and reduced input use from more efficient farming; more efficient farm management with improved communication possibilities and higher quality of work with machine-guided systems. The implementation of precision farming concepts may mitigate production risks because inputs are applied only where they are needed. While risk mitigation with precision farming is intuitive, the implementation of precision farming typically requires substantial investments that may increase financial risk (Lowenberg-DeBoer 1999). Investments in precision farming are further associated with irreversibility of the capital cost, which should be taken into account where appropriate. Farmers might prefer to wait for better information on the costs and benefits of the new technology before investing in precision farming technologies (Tozer 2009).

While the costs of precision farming technologies can, in many cases, be estimated precisely, it is more challenging to evaluate the response of the system to improved management. Production function analysis can be applied to assess the effects on crop yield and changes in input use from site-specific nitrogen management (Bachmaier and Gandorfer 2009; Meyer-Aurich et al. 2010b; Bullock et al. 2002, Rogers et al. 2016). The analysis of production functions helps us to gain insight into input–output relationships to enable the choice of the optimal rate of input as a function of the price of input and output. Site-specific production functions can be estimated, for instance, from field trial data (e.g. nitrogen rate trials) or from data simulated with crop growth models.

In rain-fed plant production systems, the response to inputs varies substantially from year to year, which complicates the determination of economically optimal rates of input. However, from an ex-post perspective, the analysis of production functions helps us to understand the crop response to inputs, and can be used to identify the economic potential of site-specific fertilization strategies. In such ex-post analyses, it is typically assumed that economically optimal rates of input are applied in both site-specific and uniform management. Therefore, it is important to consider that the results show the theoretical ex-post economic potential of site-specific management. In practical agriculture, it is not possible to determine ex-ante exact economic optimum input rates especially for nitrogen because of unpredictable weather events. This is true for site-specific management as well as for uniform management. Thus, there will always be a difference between the theoretically optimal fertiliser rate and the realized fertiliser rate. Whether the theoretically optimal fertiliser rate can be realised in practice depends on the applied site-specific technology, implemented decision algorithms and other factors, such as a uniform reference system.

Bachmaier and Gandorfer (2009) presented an approach based on production function analysis to test if there is a significant difference between site-specific economically optimum nitrogen rates. Significant differences in economically optimum input rates are a prerequisite for the profitability of site-specific fertiliser management. However, it is important to recognize that yield heterogeneity does not necessarily lead to significant differences in site-specific economic fertiliser rates (Bachmaier and Gandorfer 2009). From an economic point of view, it is important that site-specific production functions have different slopes, which cause different marginal yield responses to an additional unit of input. Thus, yield heterogeneity identified, for instance, with yield maps from combine harvesters cannot serve as a robust indicator for the profitability of site-specific fertiliser management. In this context, Rogers et al. (2016) have suggested a new metric to describe the flatness of site-specific payoff functions in order to estimate better the economic potential of site-specific input management at the field level. The metric is called relative curvature, and the authors found that the metric could help to identify field heterogeneity from an economic perspective. Relative curvature "is obtained by calculating the area lying between the graph of the pay-off function and a horizontal line that is tangent to this graph at the point of maximum pay-off (profit) over a given range of input values" (Rogers et al. 2016, p. 111).

An alternative way to assess the economic benefits of site-specific management approaches (e.g. a commercially available sensor system for nitrogen fertilization) compared to uniform management is to conduct field trials (e.g. strip trials) where different systems are tested and compared. In such trials, uniform management is often defined as farmers' usual practice. This is important to consider when such results are discussed in comparison to the potential analysis based on site-specific production functions described previously.

3.3 Analysis of Studies

Various studies have shown mixed results of the profitability of site-specific management. Lambert and Lowenberg-DeBoer (2000) reviewed 108 studies on the economics of site-specific management strategies. Of the 34 studies that deal with site-specific fertilization, 65% showed positive economic effects, 18% showed negative effects and 17% of the studies reviewed described mixed results (see

Lambert and Lowenberg-DeBoer 2000, p. 14). Bongiovanni and Lowenberg-DeBoer (2004) provided an extensive review of precision agriculture studies to analyse the potential contribution of precision agriculture technologies to a more sustainable agricultural production system. They concluded that site-specific management of inputs, like fertilisers and chemicals, reduce negative impacts on the environment. However, their case study in Argentina showed that the profitability is only modest compared to whole field management. Also, Diacono et al. (2013) concluded from a review of studies about site-specific nitrogen fertilisation of wheat that these approaches do not necessarily lead to economic advantages. In several more recent studies, the economic potential of site-specific fertiliser management is analysed both theoretically and empirically based on improved technological possibilities, which will be discussed in detail in the following sections.

3.3.1 Site-Specific Nitrogen Fertilization

Schneider and Wagner (2008) investigated the economics of site-specific fertilization strategies. They compared a sensor and a mapping approach for site-specific nitrogen fertilization in winter wheat and canola. Based on a series of eight experiments, the partial budgeting of the mapping approach resulted in an average negative contribution to profit (−14 € ha^{-1}), whereas the sensor approach provided a positive contribution to profit (16 € ha^{-1}) (Schneider and Wagner 2008, p. 419). These values do not include costs for the sensor technologies necessary for information gathering and variable-rate application. The per ha cost assumptions for the sensor approach depend on the acreage on which the technology is used, ranging from about 6 to 65 € ha^{-1} for the use on 1000 to 100 ha, respectively (Schneider and Wagner 2008, p. 426). A more sophisticated and information-intensive site-specific fertilization approach based on a neural network and decision tree algorithms resulted in the highest net profitability when compared to other fertiliser management systems. The economic advantage of this approach (partial budget) was 46 € ha^{-1} compared to uniform management, and 29 € ha^{-1} compared to the sensor approach (Schneider and Wagner 2008, p. 419). These results are in line with the findings from theoretical work by Bullock et al. (2002), who found increasing marginal profits of site-specific nitrogen fertiliser management with increasing information. However, the maximum gross economic effect was only about 7 US$ ha^{-1} when costs for information gathering were excluded (Bullock et al. 2002). Another study investigated the economic return of site-specific fertilization of nitrogen and phosphorus in Western Australia (Lawes and Robertson 2011). They found that site-specific fertiliser management provided economic benefits on six of the 20 fields investigated with an average of 15 AU$ ha^{-1} (ca. 11 € ha^{-1}), however, costs for information gathering and variable-rate application were again excluded. Lawes and Robertson (2011) also addressed the question of to what extent an increase in the number of management zones can contribute to higher economic returns. They found diminishing marginal returns with increasing number of management zones, which provides an argument for the importance to address the major managing zones. This conclusion is in contrast to the results of Schneider and Wagner (2008) and Bullock et al. (2002) discussed above. Therefore, from an economic point of view, it remains an open question as to how precise (e.g. the number of different management zones) site-specific fertiliser management should be.

It is further notable that studies based on field trials often show higher economic benefits of site-specific farming than the theoretical potential benefits derived from production function analysis (Silva et al. 2007, Meyer-Aurich et al. 2008). This is somewhat surprising, but can be explained with the reference (uniform management) chosen for the economic comparison. For example, if a uniform management system is compared with a site-specific management system, both systems rely on different sources of information. For example, while a site-specific management system uses a sensor, a uniform management system might rely on expert knowledge. If the uniform management is performed badly, the difference in the economic performance of the systems compared is higher. It can be assumed further that the implementation of site-specific fertilization contributes not only to a better consideration of production factors like fertilisers, but also to better management in general. In the analytical ex-post analyses, it is difficult, if not impossible, to distinguish between both effects. Thus, comparisons of site-specific and uniform management based on field experiments should be analysed with care.

The rather low economic advantages question the site-specific management of fertilisers from an economic point of view, which is in line with the conclusions of Oleson et al. (2004), and Liu et al. (2006). Based on payoff function analysis, Pannell (2006, p. 553) concluded that: "the benefits of using 'precision farming' technologies to adjust production input levels are often low". This conclusion results from the insight that payoff functions are often flat in the area of the economic optimum input level and, therefore, deviations from the economically optimum input level are in many cases associated with only marginal economic losses (Pannell 2006). Figure 3.1 illustrates an example of a flat payoff function and shows that, for instance, a deviation of 20% from the economic optimum input rate reduces the payoff only marginally.

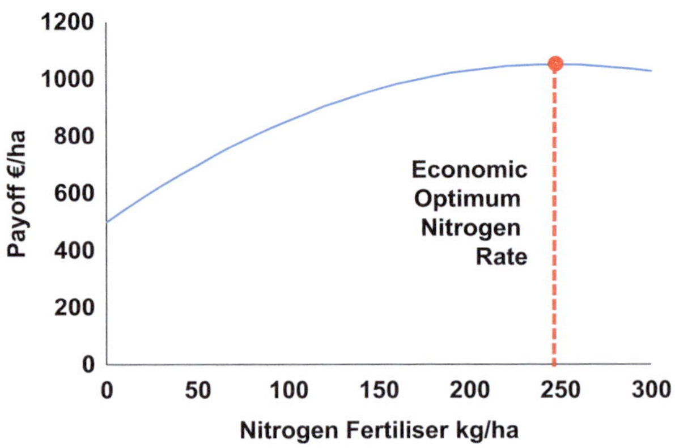

Fig. 3.1 Payoff for winter wheat as a function of nitrogen fertiliser rate (crop quality not considered). Note: The figure is based on a published nitrogen production function for the winter wheat cultivar Contur (Meyer-Aurich et al. 2010b) (Source: own illustration)

Site-specific fertiliser management, however, could result in a higher economic advantage if farmers were faced with environmental restrictions or had to internalize the environmental damage costs of fertilization. In this context Gandorfer et al. (2003) showed, for instance, that site-specific nitrogen management leads to lower abatement costs compared to uniform management when environmental targets (e.g. nitrate concentration in seepage water) have to be met. Also Rogers et al. (2016) conclude that if farmers must internalize negative external effects of sub-optimal fertiliser application, the importance of identifying economic optimum input levels increases and therefore, the economic benefit of site-specific fertiliser management.

3.3.2 *Site-Specific Management with Respect to Crop Quality*

An additional increase in benefit of site-specific fertilization may be realized if higher crop qualities can be assured and thus, the crop can be sold at higher prices. In this situation, the payoff function jumps to a higher level which may result in higher profit margins. This can be achieved by site-specific nitrogen management in wheat production considering site-specific protein functions or by quality specific harvest.

3.3.2.1 Site-Specific Nitrogen Management with Respect to Protein Concentration

Gandorfer and Rajsic (2008) provided an empirical example for such a situation where the payoff function jumps to a higher level when a specific protein concentration threshold is met and, therefore, the winter wheat price increases (Fig. 3.3). The analysis is based upon estimated winter wheat yield and protein response functions to nitrogen fertilisation (Fig. 3.2) for two experimental sites,- Wolfsdorf and Betzendorf (Bavaria, Germany). The experimental field in Wolfsdorf shows a higher yield potential because of better growing conditions in terms of precipitation, average temperature and soil conditions (Gandorfer and Rajsic 2008).

The extent of the jumps in the payoff function depends on the underlying yield and protein response functions, but also on the protein premium schemes. Because protein premium schemes differ both from year to year and between crops, the economic benefits of accounting for crop quality in terms of protein concentration also vary from year to year. For illustration, Fig. 3.4 shows producer prices for different quality grades of winter wheat (A, B and Feed Quality) in terms of protein concentration. To be graded as “A-Quality’ wheat, the protein concentration must be above 13.5%. Wheat with a protein concentration in the range between 12% and 13.5% falls into the “B-Quality’ category. Clearly, there are years in which a high protein concentration is beneficial (e.g. 2010) and years with marginal price differences only among different qualities (e.g. 2012).

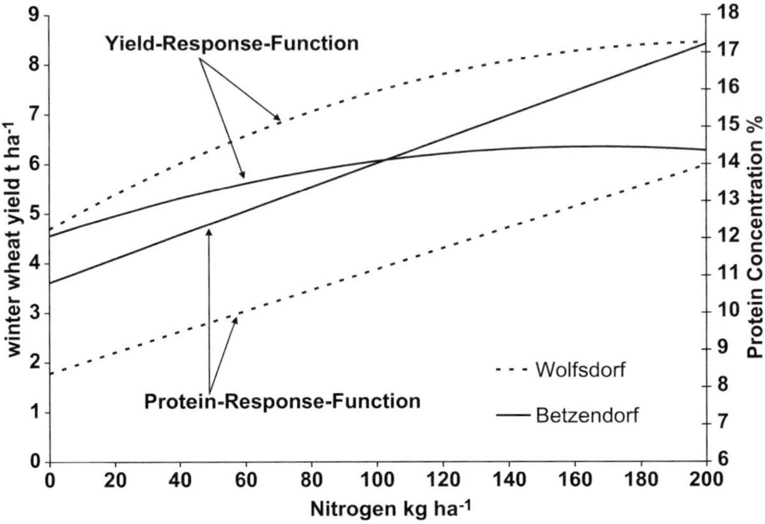

Fig. 3.2 Estimated average (2000–2002) winter wheat yield and protein response functions to nitrogen fertilisation (Source: Gandorfer and Rajsic 2008)

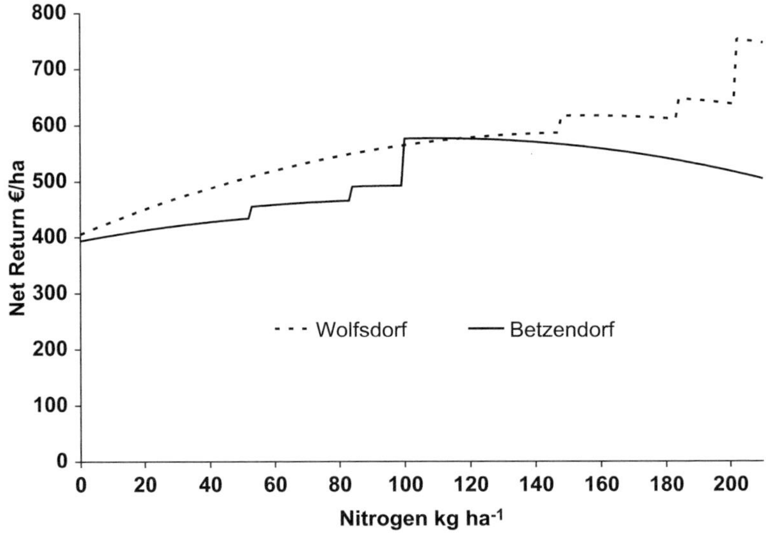

Fig. 3.3 Average net return function calculated for price regimes from 2004 and 2005 for the experimental sites in Wolfsdorf and Betzendorf. Net return is defined as crop revenue minus nitrogen fertiliser cost (Source: Gandorfer and Rajsic 2008)

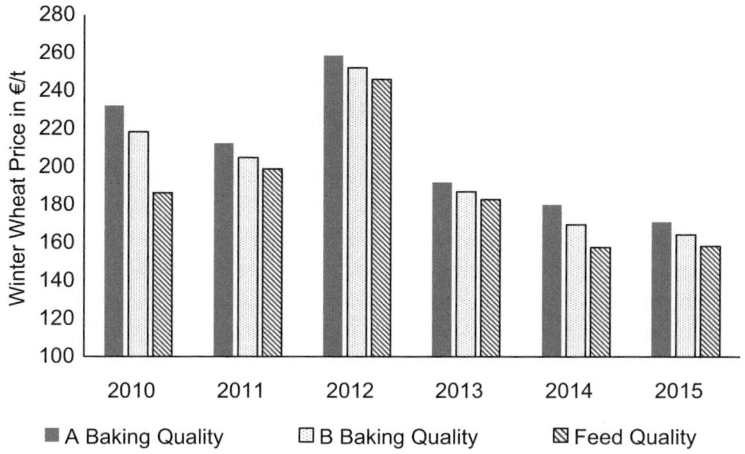

Fig. 3.4 Protein premium schedules for Bavaria, Germany, Source LfL (2016)

Meyer-Aurich et al. (2010a) provided an economic analysis of site-specific fertiliser strategies with consideration of crop quality based on data from an on-farm field experiment. In their study they proposed a spatial econometric approach to analyse crop yield and quality response to nitrogen fertiliser to improve nitrogen management. However, they did not find a clear economic advantage of site-specific fertilization when crop quality was considered in terms of the German protein premium scheme. The gross economic potential of site-specific nitrogen management with respect to protein concentration was estimated to be 2.57 € ha^{-1} only without considering the fixed costs associated with the site-specific fertiliser application approach.

3.3.2.2 Site-Specific Harvest Management (Grain Segregation by Protein Concentration)

The economic effects of grain segregation and blending by protein concentration has been addressed by various authors in the past (e.g. Sivaraman et al. 2002). These analyses were performed either at the level of the grain elevator or at later stages in the grain value chain. New precision farming technologies now make it possible to realize site-specific harvest management with the idea of shifting back the economic benefits of grain segregation and blending from the grain elevator to the farm operations. Thus, several authors have studied the economic effects of various approaches of site-specific harvest management and grain segregation recently (e.g. zone harvesting or separation in harvester) (Tozer and Isbister 2007; Meyer-Aurich et al. 2010b; Martin et al. 2013).

Meyer-Aurich et al. (2010b) discussed that 'on the go' sensors could help to separate grain quality during harvest, and different fractions could be sold at different prices. In contrast to the site-specific fertiliser strategy, this strategy may have a higher economic effect, especially if the necessary crop quality cannot be achieved for the whole field and price incentives for higher grain qualities are set. Therefore, the positive economic effect of site-specific harvest management is based on a higher average crop price compared to whole field harvest. Because a higher average crop price, site-specifc harvest management can shift the site-specific payoff function to a higher level. Martin et al. (2013) have identified three important variables that determine the economic benefits of grain segregation at the field level. In addition to the average protein concentration of the field (1) and within-field variability of protein concentration (2), the price premiums for protein (3) are relevant.

Meyer-Aurich et al. (2010b) constructed model calculations based on fertiliser response experiments that show the possible crop yield and grain quality response of wheat to nitrogen fertiliser supply. Based on virtual fields with heterogeneous response, the economic gross benefit of site-specific harvest management resulted in an advantage ranging from −2 € ha^{-1} to 33 € ha^{-1}. Although the relative profitability of site-specific harvest management is limited, it can have a risk reducing effect. This is demonstrated by the example shown in Fig. 3.5 for two price scenarios for baking quality wheat.

With uniform harvest (solid line), the highest net returns (above fertiliser cost) can be obtained with a fertiliser rate of about 170 kg N per ha, which is about 80 € ha^{-1} higher than the maximum net return for feed quality at a premium for baking quality of 0.9 € per ton wheat (top graph in Fig. 3.5). This premium was the average premium received by Bavarian farmers from 2009 to 2016. At a lower protein premium (bottom graph in Fig. 3.5), returns above fertiliser costs are reduced accordingly. Since parts of the field achieve baking quality at N rates lower than 170 kg ha^{-1}, at these fertiliser rates the possibility of separating different qualities can generate a higher profit compared to a uniform harvest by selling a fraction of the harvest as quality wheat. This advantage is illustrated with the dotted line in Fig. 3.5. Even though the maximum net return above fertiliser cost with site-specific harvest management does not exceed the maximum of the net return with uniform harvest, the window of fertiliser levels that result in higher net returns is substantially bigger. In other words, within a window of nitrogen rates from 158 and 179 kg N ha^{-1}, net returns are higher with site-specific harvesting because within this range baking quality can be achieved in one of the modelled parts of the field only. The separation of the higher quality grains results in a higher economic return for this part of the grains and averaged over the whole field (dotted line). Without grain separation, all grains are assumed to be sold at a lower price since the average protein content is below the threshold.

The results indicate that separating different grain qualities during harvest can assure high profits, even when the protein requirement is not achieved for the whole field. This may reduce the producer's risk, i.e. failure to achieve the required protein quality in the whole field.

Tozer and Isbister (2007) evaluated the economic benefits of harvesting by management zones, and identified situations in terms of field layout, and yield or quality scenarios where site-specific harvest management can generate economic benefits. The economic effects of harvesting by management zones ranged from −8 AU\$ ha^{-1} to 30 AU\$ ha^{-1} (ca. –6–20 € ha^{-1}) for the different scenarios analysed (Tozer and Isbister 2007, p. 158). The assessment included additional costs distances trav-

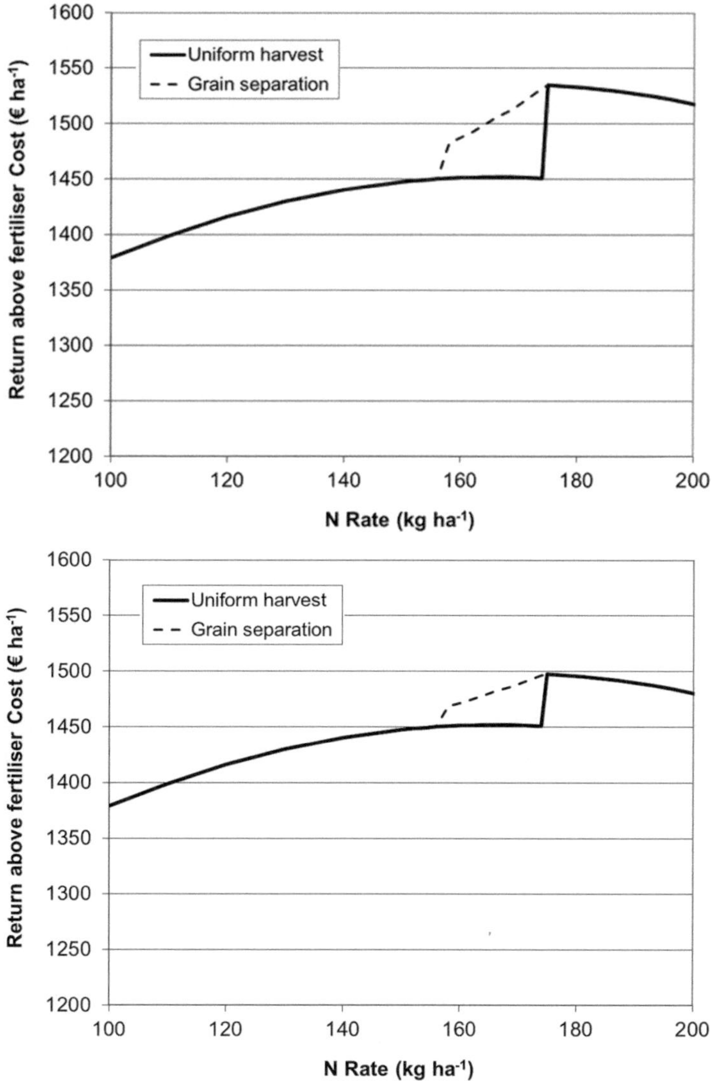

Fig. 3.5 Returns above fertiliser costs with uniform and separate harvest management with premiums for baking quality wheat of 0.9 € (top graph) and 0.5 € (bottom graph) per ton of wheat (Based on model calculations in Meyer-Aurich et al. 2010b)

elled that arise from harvesting by management zone. The authors showed that because of the underlying protein premium schedules and additional harvesting costs, blending the grain from the whole field can lead to a higher gross crop revenue compared to harvesting by management zone.

A limitation of available studies is that they often do not account for the total cost of site-specific harvest management including technology costs for grain segregation and additional storage and logistic costs. Particularly, additional storage and logistic costs may be high, and can diminish the economic benefits of site-specific harvest management. The impact of site-specific management approaches that consider grain quality (including separate harvest) on the environmental has not yet been studied sufficiently. While it is intuitive to assume that site-specific management can save on unnecessary amounts of fertiliser where they are not needed, site-specific management could also enable the exploitation of economic potentials leading to negative environmental effects.

3.4 Conclusions

Economic benefits of site-specific fertiliser management are often limited because of flat site-specific payoff functions in the area where the economic optimum is located. Even though information technologies are expected to become less cost-intensive over time, this will not overcome the general limitation of flat site-specific payoff functions. Furthermore, the necessary sensor technologies and advanced site-specific application technologies may not necessarily become cheaper in the future. Therefore, from an economic point of view future potentials of precision farming are, particularly given for technologies which generate a new payoff function, at a higher level rather than technologies which aim at improving management decisions (see also Gandorfer et al. 2011, Pannell 2006). One example of such a technology might be site-specific harvest management. However, available studies currently do not prove substantial economic advantages of site-specific harvest management, but do show a potential risk-reducing effect. Improved efficiency in agricultural systems with precision farming may provide environmental benefits. Further research is required to provide an economic assessment of this potential positive externality. The advantage of site-specific fertiliser management and harvest management may be higher if farmers were faced with environmental restrictions or in a situation where the costs of environmental damage from fertiliser use must be accounted for.

References

Bachmaier M, Gandorfer M (2009) A conceptual framework for judging the precision agriculture hypothesis with regard to site-specific nitrogen application. Precis Agric 10(2):95–110

Bongiovanni R, Lowenberg-DeBoer J (2004) Precision agriculture and sustainability. Precis Agric 5(4):359–387

Bullock DS, Lowenberg-DeBoer J, Swinton SM (2002) Adding value to spatially managed inputs by understanding site-specific yield response. Agric Econ 27(3):233–245

Diacono M, Rubino P, Montemurro F (2013) Precision nitrogen management of wheat. A review. Agron Sustain Dev 33(1):219–241

Gandorfer M, Rajsic P (2008) Modeling economic optimum nitrogen rates for winter wheat when inputs affect yield and output-price. Agric Econ Rev 9(2):54–64

Gandorfer M, Meyer-Aurich A, Heißenhuber A (2003) An economic assessment of site-specific and homogenous nitrate fertilization management in consideration of environmental restriction. Programme book of the joint conference of ECPA-ECPLF, Berlin, pp 407–408

Gandorfer M, Meyer-Aurich A, Kapfer M (2011) Ökonomische Aspekte von Precision Farming - aktueller Stand & Perspektiven. In: Hambrusch J, Larcher M, Oedl-Wieser T (Hrsg) Jahrbuch der Österreichischen Gesellschaft für Agrarökonomie, Band 20(2):35–44

Griffin TW, Lowenberg-DeBoer J, Lambert DM, Peone J, Payne T, Daberkow SG (2004) Adoption, Profitability, and Making Better Use of Precision Farming Data, Staff Paper #04–06. Department of Agricultural Economics, Purdue University, USA, West Lafayette

Khanna M, Epouhe OF, Hornbaker R (1999) Site-specific crop management: adoption patterns and incentives. Rev Agric Econ 21(2):455–472

Lambert DM, Lowenberg-DeBoer J (2000) Precision agriculture profitability review. West Lafayette, Site Specific Management Center, School of Agriculture, Purdue University. Available at: https://www.agriculture.purdue.edu/ssmc/FRames/newsoils.doc. Accessed 09 Sept 2016

Landesanstalt für Landwirtschaft (LfL) (2016) Deckungsbeiträge und Kalkulationsdaten. Bayerische Landesanstalt für Landwirtschaft. München. Available at: https://www.stmelf.bayern.de/idb. Accessed 24 Aug 2016

Lawes RA, Robertson MJ (2011) Whole farm implications on the application of variable rate technology to every cropped field. Field Crop Res 124(2):142–148

Liu Y, Swinton SM, Miller NR (2006) Is site-specific yield response consistent over time? Does it pay? Am J Agric Econ 88(2):471–483

Lowenberg-DeBoer J (1999) Risk management potential of precision farming technologies. J Agric Appl Econ 31(2):275–285

Martin CT, McCallum JD, Long D (2013) A web-based calculator for estimating the profit potential of grain segregation by protein concentration. Agron J 105(3):721–726

Meyer-Aurich A, Gandorfer M, Heißenhuber A (2008) Economic analysis of precision farming technologies at the farm level: two German case studies. In: Castalonge OW (ed) Agricultural systems: economics, technology, and diversity. Nova Science Publishers, Hauppage, pp 67–76

Meyer-Aurich A, Griffin TW, Herbst R, Giebel A, Muhammad N (2010a) Spatial econometric analysis of a field-scale site-specific nitrogen fertilizer experiment on wheat (*Triticum aestuvum L.*) yield and quality. Comput Electron Agric 74(1):73–79

Meyer-Aurich A, Weersink A, Gandorfer M, Wagner P (2010b) Optimal site-specific fertilization and harvesting strategies with respect to crop yield and quality response to nitrogen. Agric Syst 103(7):478–485

Oleson JE, Sorensen P, Thomson IK, Erikson J, Thomsen AG, Berntsen J (2004) Integrated Nitrogen input systems in Denmark. In: Mosier AR, Syers JK, Freney JR (eds) Agriculture and the nitrogen cycle. Island Press, Washington, DC/Covelo/London, pp 129–140

Organisation for Economic Co-operation and Development (OECD) (2016) Farm management practices to foster green growth, OECD Green Growth Studies. OECD Publishing, Paris. Available at: https://doi.org/10.1787/9789264238657-en Accessed 09 Sept 2016

Pannell DJ (2006) Flat earth economics: the far-reaching consequences of flat payoff functions in economic decision making. Rev Agric Econ 28(4):553–566

Reichardt M, Jürgens C, Klöble U, Hüter J, Moser K (2009) Dissemination of precision farming in Germany: acceptance, adoption, obstacles, knowledge transfer and training activities. Precis Agric 10(6):525–545

Robertson M, Llewellyn R, Mandel R, Lawes R, Bramley R, Swift L, Metz N, O'Callaghan C (2012) Adoption of variable rate fertiliser application in the Australian grains industry: status, issues and prospects. Precis Agric 13(2):181–199

Rogers A, Ancev T, Whelan B (2016) Flat earth economics and site-specific crop management: how flat is flat? Precis Agric 17(1):108–120

Schneider M, Wagner P (2008) Ökonomische Effekte der teilflächenspezifischen Bewirtschaftung auf betrieblicher Ebene. In: Werner A, Dreger F, Schwarz J. Informationsgeleitete Pflanzenproduktion mit Precision Farming als zentrale inhaltliche und technische Voraussetzung für eine nachhaltige Entwicklung der landwirtschaftlichen Landnutzung – preagro II. Available at: http://preagro.auf.uni-rostock.de/Veroeff/preagro_Abschlussbericht_2008.pdf. Accessed 25 Aug 2016

Silva CB, Do Vale SMLR, Pinto FAC, Müller CAS, Moura AD (2007) The economic feasibility of precision agriculture in Mato Grosso do Sul State, Brazil: a case study. Precis Agric 8(6):255–265

Sivaraman E, Lyford CP, Brorsen BW (2002) A general framework for grain blending and segregation. J Agribusiness 20(2):155–161

Tozer PR (2009) Uncertainty and investment in precision agriculture – Is it worth the money? Agric Syst 100(1–3):80–87

Tozer PR, Isbister BJ (2007) Is it economically feasible to harvest by management zone? Precis Agric 8(3):151–159

Zarco-Tejada P, Hubbard N, Loidjani P (2014) Precision Agriculture: An Opportunity for EU Farmers – Potential Support with the CAP 2014–2020, Policy Department, Structural and Cohesion Policies, Agriculture and Rural Development, European Parliament. Available at: http://www.europarl.europa.eu/RegData/etudes/note/join/2014/529049/IPOL-AGRI_NT%282014%29529049_EN.pdf. Accessed 09 Sept 2016

Chapter 4
Economics of Site-Specific and Variable-Dose Herbicide Application

Jens Erik Ørum, Per Kudsk, and Peter Kryger Jensen

Abstract Site-specific application of pesticides has so far focused mainly on herbicides. The purpose of precision farming technologies in relation to herbicide use is to reduce herbicide cost and environmental impact from spraying, but at the same time to achieve acceptable weed control. Another purpose is to increase the spraying capacity, to reduce the number of sprayer refills, and finally to minimize time spent on weed monitoring. In this chapter the relevance and profitability of four precision herbicide application technologies, two weed detection technologies and a low dose decision support system (DSS) is analysed. With a low dose herbicide, cost can be reduced by 20–50%. It requires, however, proper monitoring of weeds, which can be a time-consuming task that again requires that the farmer is able to identify the dominant weed species. The current development of high-speed camera and software systems can help to detect and map individual weeds, and some systems have proved to be cost effective for certain weeds.

Keywords Weed detection • Weed control • Crop protection • Low dose DSS • Precision spraying

4.1 Introduction

Research into site-specific application of pesticides has focused mainly on herbicides, although some research has been carried out within the area of disease detection and variable application of fungicides (Pedersen 2003). Preventive and site-specific treatment with insecticides is complex because insects are difficult to

J.E. Ørum (✉)
Department of Food and Resource Economics, University of Copenhagen, Copenhagen, Denmark
e-mail: je@ifro.ku.dk

P. Kudsk • P.K. Jensen
Department of Agroecology, Aarhus University, Aarhus, Denmark
e-mail: per.kudsk@agro.au.dk; pkj@agro.au.dk

S.M. Pedersen, K.M. Lind (eds.), *Precision Agriculture: Technology and Economic Perspectives*, Progress in Precision Agriculture,
https://doi.org/10.1007/978-3-319-68715-5_4

monitor in the field. In contrast, weeds tend to have a patchy distribution in the field and some diseases also typically occur in the same areas in the field with specific micro climatic conditions, although weed distribution cannot be considered invariable. On the other hand, farmers are aware of the potential savings of chemicals through precision spraying, although they may also have reservations about the potential benefits because of technical difficulties and lack of decision support systems when using site-specific technologies (e.g. Pedersen et al. 2004).

Examples of precision technologies relevant for site-specific herbicide application are:

- High precision cell and micro spraying
- Boom section and nozzle control
- Boom section control to reduce overlap
- Direct injection
- Real time weed detection with high speed cameras
- DSS systems for low dose herbicide use recommendations

The purpose of these technologies is to determine what, where and when to control weeds efficiently preferably at a low cost. Farmers' incentives for using these technologies are to (1) increase the spraying capacity and efficiency (more acreage covered per time unit, reduce number of sprayer refills), (2) minimize time-consuming activities like manual weed monitoring and (3) have access to decision support on herbicide choice and dose to reduce herbicide cost without loss of efficacy on the weeds. The incentive of the society to support the adoption of these technologies is to reduce the unintended environmental impact from spraying and at the same time maintain effective and competitive food production.

Site-specific application is about where to spray, while variable-dose application is about how much to spray. Knowledge of weed distribution in time and space is essential for both approaches. The main difference is that understanding weed response to herbicides, crop and weed competition and weed population dynamic is a prerequisite for variable-dose application.

High precision spraying of individual weed plants with broad-spectrum herbicides like glyphosate is the ultimate example of site-specific herbicide application. The system must distinguish between crop and weed. What isn't a crop is a weed plant, or everything green outside the crop row or GPS located crop plant is a weed. Site-specific solutions focus mostly on technology and engineering and less on biology and agronomy. Depending on the weed species and density, site-specific spraying could ideally reduce herbicide use by a factor of 100 or 1000 compared to a full dose spraying of the whole field.

Assuming a density of 300 weeds per m^2, Mathiassen et al. (2016) found that glyphosate doses applied at an early growth stage could be reduced from currently recommended glyphosate dose of 560–720 g ha^{-1} to 2.5–32.0 g ha^{-1} by using a Drop on Demand inkjet printer application system (DOD). Thus, specifically targeting the weed plants with the DOD system or similar devices may lead to significant reductions (factor 20–300) in herbicide field rates.

Weed detection, planning and spraying are the three dimensions in chemical weed control with variable-doses, and in this chapter we will evaluate the development and economic potential of some of the technologies above listed relating to these three dimensions. This evaluation will first of all be based on studies in the literature. In the case of automated weed detection and DSS systems for low dose herbicide recommendations we will, however, give an example of a potential 40% herbicide use reduction with a low dose DSS system that has not been exploited because of a lack of weed monitoring capacity, lack of economic incentive and a too great efficiency of pre-emergence herbicides. Profitability of investments in precision technology and future herbicide cost reductions will be calculated with a 4% discount rate and a 5 year lifetime of the investment, resulting in a capitalization factor of 4.6.

4.2 Application Technologies

4.2.1 Site-Specific Application and Variable-Dose

Variable-dose application can be achieved in many ways, from varying the application by turning on and off the sprayer while driving, adjustment of speed and tank pressure or by advanced high precision control of individual spray boom sections and nozzles (see e.g. Grisso et al. 2011). Conventional boom sprayers are usually mounted with a water tank with chemicals that are mixed with water. Usually 2–3 different chemicals are mixed at a time for each treatment. To conduct variable-dose application, some sprayers are equipped with devices to regulate the amount of chemicals simultaneously with the spraying. A GPS receiver and a tractor computer can be installed to regulate and carry out variable-dose application of pesticides. Conventional boom sprayers can also be divided into different sections to enable site-specific application along the boom. One example is described in the Sensispray development project, in which a 27-m length boom sprayer with seven sections was equipped with sensors to control spray volume per boom section, thereby having a length of about 3–4.5 m for each section (van de Zande et al. 2009).

4.2.2 Micro and "Cell" Spraying

As stated in Franco et al. (2017) the ultimate site-specific weed management strategy would be to apply one drop of herbicide per weed plant. This strategy was tested by Lund et al. (2006), with micro spray tubes that open and close individually with solenoid valves. In a field with 100 weed plants per m^2 and 20 tubes per 100 mm, it was possible to obtain 84% weed control with as little as 27 g of glyphosate ha^{-1}. Lund et al. (2008) has shown that with vision based spraying techniques one can treat areas of 100 × 100 mm individually (named 'cell spraying' by the authors).

This approach could potentially reduce pesticide use by 50–70% compared to conventional boom spraying.

Compared to a full dose of conventional herbicides the micro spraying system offers a potential saving of 10–20 € ha^{-1} and the cell spraying system offers a saving of 5–14 € ha^{-1} (Franco et al. 2017). The main question is, however, whether the cost reductions can justify the investment in a system with the required remote controlled micro tubes, image or video devices for real-time detection of weeds or alternatively high precision weed maps in combination with RTK-GPS technology etc. On the other hand, the capacity of high precision systems is another challenge for their commercialization. Despite the low capacity, a small micro sprayer with a 1-m boom could be sufficient if the system is unmanned and RTK GPS guided. High precision micro tube systems are not yet available and affordable for farmers, and the video guided cell spraying systems are commercially available for row crops only (Franco et al. 2017).

4.2.3 *Boom Section and Nozzle Control*

A boom sprayer (Fig. 4.1) is usually supposed to apply exactly the same dose in its full length. There are, however, at lot of methods to control and adjust the dose and to open and close parts of the sprayer to produce a variable-dose or site-specific pesticide application.

The boom consists of sections (H) with several nozzles per section (I) that, on advanced sprayers, can be opened or closed by the operator or automatically according to a plan and a GPS position (F). The dose is determined by nozzle size, spray pressure and driving speed. With more advanced sprayers, the dose is controlled

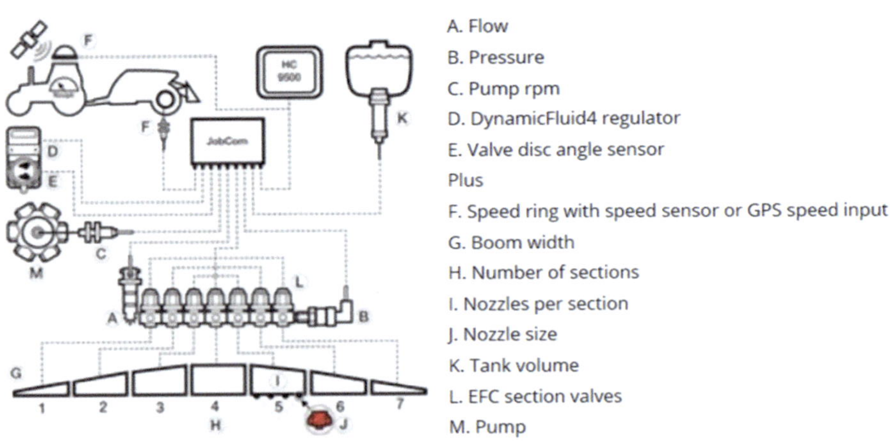

Fig. 4.1 Schematic overview of a boom sprayer (DynamicFluid4 sprayer, Hardi International A/S)

with a separate pump (M) that can be controlled by a computer that activates specific boom sections and nozzles (F). Using boom section and nozzle control units together with a crop scanner or GPS with RTK positioning, it is possible to vary the dose and do site-specific spraying applications. The challenge is, however, to identify and exploit the most profitable opportunities.

In cereal crops, high precision and reduced herbicide consumption could be achieved with traditional boom sprayers equipped with remote control of the individual nozzles or boom sections. With a traditional boom sprayer with between 2 and 4 nozzles per m boom spraying precision with remote control of every nozzle would be 250–500 mm. If we consider a weed density that varies from 25 to 200 plants per m^2 and a spraying precision (wideness of the sprayed area per nozzle) between 250 and 500 mm, then it would almost certainly result in full spraying of the field. A precision of 30–100 mm would be needed to reduce the pesticide use significantly (Franco et al. 2017). However, in that case, the cost savings of reduced herbicide would probably not be sufficient to pay for video devices, weed mapping, RTK-GPS equipment and remote controlled solenoid valves for individual nozzles or boom sections. To make a precision of 250–500 mm spraying profitable, a very patchy variation in the weed density and distribution would be required, such as for thistles and couch grass, which tend to grow in colonies, patches, spots and clusters.

A study by Franco et al. (2017) showed that the profitability of increased precision in glyphosate spraying of thistle patches in cereals decreases significantly with an increasing precision. The marginal value (v) of an extra unit was found to be a power function of length of the boom (l), the length of the controlled units, either nozzles or boom sections, (w) and the cost of a full herbicide dose (H), such that:

$$v = H\alpha\beta\left(\frac{l}{w}\right)^{\beta-1}.$$

The α and β values depend on the size and distribution of the weed patches. Below is an example with a 40-m boom, many and scattered patches of weeds (α=0.56 and β=-0.73) and few and concentrated patches of weeds (α=0.29 and β=−1.01). Figure 4.2 shows how the reduction in marginal costs decreases rapidly with an increased precision from an increased number of remotely controlled units (RC).

Figure 4.2 shows that the first step from broadcast spraying to a simple (1 unit) open and closing of the whole sprayer reduces the herbicide cost by 6–9 € ha^{-1}. With an extra control unit, the 40-m sprayer is divided into two 20-m sections, reducing the herbicide cost by an extra 2–3 € ha^{-1}. With 5–6 units corresponding to 5–10-m sections the extra herbicide cost reduction is marginal. In this case a five- or 10-m precision is cost efficient, but this knowledge could be derived only by using a very detailed weed map that shows the thistle patches.

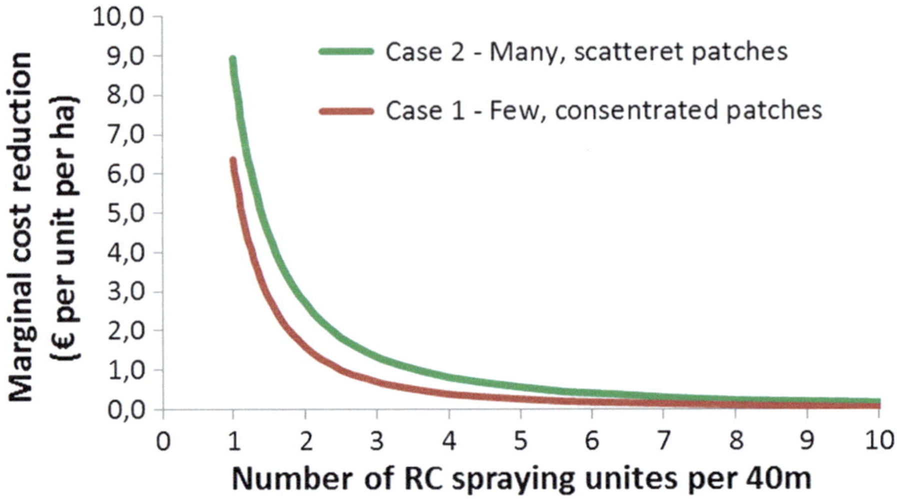

Fig. 4.2 Marginal cost reduction (€ha^{-1})per an extra controlled unit (Franco et al. 2017)

4.2.4 Boom Section Control and Reduced Overlap

Reducing overlap when spraying is a potential source for cost reduction. A rough estimate of a 5% reduction in pesticide use could be achieved by reducing overlapping when spraying along field edges (Petersen et al. 2017). It takes automated boom section control, GPS and software to realize this reduction. Most new sprayers are manufactured with electronic on/off switches for individual boom sections. To control the individual sections computer and software such as Rinex AS 7500 (by Leica Geosystems), GPSswitch (by Amazone), Swath Control Pro (by John Deere), AgGPS FieldManager (by Trimble) etc. are needed.

According to a Danish sprayer retailer (Mertz, 2017), 80% of all new sprayers are equipped with automated section control. Computer software and installation may sum to around 3000 €. If not already installed, a GPS unit is available at a cost of around 2000 €.

Given a 4% discount rate and a 5-year lifetime, a 3000 € investment in sprayer section control or GPS equipment must produce a yearly cost reduction that is equivalent to 649 €.

Table 4.1 provides the pesticide costs for Danish arable crops 2015 and the required area (ha) needed to make a cost reduction of pesticides.

Table 4.1 shows that a 5% pesticide cost reduction on 90 ha of winter wheat is sufficient to balance the 3000€ automated section control investment. A GPS is a requirement for automated section control and RTK may be a prerequisite for even finer control of individual nozzles (1 cm accuracy). If a 5% reduction in pesticide costs alone is needed to pay for these additional investments, an extra area of

Table 4.1 Pesticide costs (€ ha^{-1}) for Danish arable crops 2015* and required area (ha) to make a 5% cost reduction of pesticides

Crop			Costs		Technology			
	1.000 ha	Weight	Pesticides	5% reduction	Section control	GPS	RTK	Total
					Investment (€)			
					3000	2000	13,000	18,000
			Costs (€ ha^{-1})		Break even area (ha)			
Winter wheat	570	31%	145	7.3	90	60	388	538
Winter barley	110	6%	104	5.2	125	83	542	750
Spring barley	490	27%	60	3.0	215	143	933	1291
Canola	190	10%	250	12.5	52	35	225	312
Seed grass	27	2%	96	4.8	135	90	584	809
Beets	25	1%	284	14.2	46	31	198	275
Starch pot.	27	1%	430	21.5	30	20	131	181
Maize	180	10%	83	4.1	157	105	682	944
Fodder grass	200	11%	0	0.0	–	–	–	–
Weighted avg.	1800	100%	114	6	114	76	495	685

Source: Statistics Denmark (DST 2016) and MST (2017)
Note: Assumption: 5 year lifetime and a 4% discount rate

60–388 ha of winter wheat or an extra area of 76–495 ha of a standard crop rotation is needed to achieve a breakeven of these investments.

Overlapping, however, does not just occur at field edges, but also between the spray tramlines. This kind of overlapping is possibly responsible for another 5% overlap (Petersen et al. 2017). However, to reduce this overlap, opening and closing of full sections is not enough. In this case, individual control of nozzles is needed to reduce the overlap. A solution with an additional control of the outermost two nozzles and the next two nozzles is available. It also takes a higher precision to exploit this overlap. If this is not installed, the investment in RTK may increase by approximately 13,000 € per unit. Instead of very precise spraying, it may be more cost efficient to invest in more precise seeding. The RTK may still be needed but the investments in the control of individual nozzles are saved and seeds (and fertilizers) can be saved.

The benefits of auto guidance and site-specific management are further described in an IPNI guideline (IPNI 2017).

4.2.5 Direct Injection

Farmers' spraying strategies usually require the mixing of herbicides into an exact amount of water prior to spraying. Hence, they need to estimate the precise amount for each herbicide required for a particular field to avoid having to empty the tank after each operation in the field and potentially waste herbicide. This approach conflicts with the idea of precision spraying (Pedersen 2003). It is therefore vital that site-specific spraying systems (1) separate herbicides and water and (2) use efficiently the knowledge of weed susceptibility to herbicides and weed distribution for estimating the optimum herbicide mixture.

Injection sprayers provide these attributes as the various undiluted herbicides are kept in a container, separated from the water tank. Water is pumped through the nozzles and herbicides are injected into the water (Walter and Heisel 2001). Commercial injection systems usually have around five containers for different herbicides (Fig. 4.3). With the injection system there are no leftovers after spraying and there are no herbicides in the water tank. An injection system can, in principle, be mounted on any hydraulic sprayer. Although several injection systems are commercially available they need further improvement in regard to reaction time and cleaning of containers and costs are still very high (Anglund and Ayers 2003).

Fig. 4.3 Example of pesticide containers on injection sprayer (Photo: Kyndestoft Maskinfabrik ApS)

4.3 Planning and Low Dose DSS

4.3.1 CPO-Weeds

Weed detection, planning and spraying are the three dimensions in variable-dose chemical weed control. As indicated above, different spraying and application methods and strategies have different implications. The subsequent question related to the planning is which herbicides and doses to apply?

In most fields, the farmers face the challenge to control weed flora consisting of several weed species rather than just one or two species, and he/she can choose among a wide range of herbicides. Most herbicides control more than just one weed species, some weed species must be controlled more effectively than others. The effect of two or more herbicides is to some extent additive, but some herbicides cannot be applied together, and efficacy of the herbicides often depends on climatic conditions, crop density and weed growth stage. For that reason the optimal herbicide recommendation will often be a combination of smaller than recommended doses of different herbicides. A few days later, the optimal recipe may be different, e.g. due to weeds growing larger or changing climatic conditions. Many active ingredients of herbicides are sold as premixes, but often farmers are able to make up their own herbicide mixes, tailor-made for their specific weed problems and these mixtures are often cheaper. The problem is, however, to find the optimal 'recipe'. Here a decision support system like CPO-weed is relevant.

Crop Protection Online-Weeds (CPO-Weeds) is a decision support system for chemical weed control developed in Denmark and subsequently adjusted to conditions in other countries. Several trade names have been used including PC Plant Protection (Denmark), Crop Protection Online (Denmark), VIPS-Ugras (Norway), CPOWeeds (Spain) and DSSHerbicide (Poland, Germany). In each country, the CPO-versions are adjusted according to herbicide availability and parameterized accordingly, whereas the algorithms and calculations follow the same concept (Sønderskov et al. 2016).

Herbicide recommendations in CPO-Weeds are calculated through a three step process following the user's input on weed species as well as growth stage and density of each weed species in the field. The first and second step is to determine the need for control and the level of control required for each of the reported weed species, respectively. The threshold for each weed species depends on the crop, crop growth stage and crop density as well as the growth stage and density of the weed species. The most competitive and problematic weed species will be controlled effectively, whereas less competitive weed species at low densities are either tolerated or partly controlled. The assessment of need for control and required level of control is based solely on expert knowledge. The third step of the decision process is the selection of herbicide solutions including herbicide mixtures that can provide the required control of each of the weed species. Potential herbicide tank mixtures combining up to four different herbicides at specified doses are identified using the additive dose model (ADM). Finally, herbicide solutions are ranked according to either the costs, the TFI or the pesticide load using a linear programming (LP) method. The concept behind the three steps is explained in more detail in Sønderskov et al. (2016).

The development of CPO-Weeds was initiated following a political decision in Denmark to reduce pesticide use, and CPO-Weeds and field experiments in various countries have revealed the potential for reduction of between 20 and 40% compared to labelled rates or standard recommendations (Sønderskov et al. 2016).

Even though CPO-Weeds are considered robust and trustworthy by both farmers and advisors, the number of farmers subscribing to the system is relatively small. A survey (Jørgensen et al. 2007) revealed that even though most farmers expressed considerable confidence in the recommendations provided by CPO-Weeds, the DSS did not fit into their way of decision-making. The advisory service in Denmark, however, makes use of CPO-Weeds, and as many farmers receive their advice on herbicide use from the advisors, they are indirectly end users benefitting from CPO-Weeds. The survey also revealed that a limiting factor for an increased uptake of CPO-Weeds is the need for field monitoring of weeds, which can be a time-consuming task and requires high level of knowledge on weed identification that farmers often do not possess. Another limiting factor was a general lack of economic incentives for reducing herbicide doses.

4.3.2 CPO-Weed Scenarios

In CPO-Weeds all weeds are reported by name, density and growth stage, but some weeds are more important than others, in the sense that their presence will prompt higher doses or the need for more expensive herbicide. In theory, there is an unlimited number of weed scenarios in CPO-weed, but 24 typical weed scenarios related to five crops are available for batch calculations and demonstration purpose.

To give an impression of the functionality and input needed for CPO-weed recommendations and to stress that intelligent and cost effective weed control requires knowledge, we will go into some detail with the 24 weed scenarios and the corresponding CPO-weed herbicide recommendations. For simplicity, however, density thresholds and growth stages are not shown or discussed, but all weed species are treated either as pre-emergence or early post-emergence under 'normal' weather conditions.

Table 4.2 shows 23 key weed species found in the 24 weed scenarios and Table 4.3 shows the 24 weed scenarios and the required level of control for each of the key weed species.

4.3.3 Field Inspection and Weed Sampling

Weed scouting takes time and requires expertise on weed species identification at early growth stages.

Table 4.2 EPPO codes, scientific name, and English names for 23 key weed species used in the CPO-weed calculations

EPPO code	Latin name	English name
ALOMY	*Alopecurus myosuroides*	Black grass
APESV	*Apera spica-venti*	Loose silky-bent
AVEFA	*Avena fatua*	Wild oat
BRSNN	*Brassica napus L.*	Oil-seed rape (voluntieer plants)
CAPBP	*Capsella bursa-pastoris*	Shepherd's purse
CHEAL	*Chenopodium album L.*	Fat hen
CIRAR	*Cirsium arvense*	Canada thistle
AGRRE	*Elytrigia repens*	Couch-grass
CONAR	*Fallopia convolvulus*	Field bindweed
GAESS	*Galeopsis spp.*	Hemp-nettle
GALAP	*Galium aparine*	Cleavers
GERdi	*Geranium spp.*	Cranes-bill
LAMSS	*Lamium spp.*	Dead-nettle
POAAN	*Poa annua*	Annual meadow grass
POLAV	*Polygonum aviculare*	Knotgrass
POLPE	*Polygonum persicaria L.*	Redshank
SOLNI	*Solanum nigrum*	Black nightshade
STEME	*Stellaria media*	Common chickweed
TRIAE	*Tricicum aestivum L.*	Wheat (voluntieer plants)
MATIN	*Tripleurospermum perforatum*	Scentless mayweed
VERAR	*Veronica arvensis L.*	Wall speedwell
VERPE	*Veronica persica*	Common field-speedwell
VIOAR	*Viola arvensis*	Field pansy

The CPO-weed holds a manual for weed scouting. It is recommends that five representative samples (50 × 50 cm) are taken per field to calculate the average weed density. In the case of large patches with a significantly higher weed density, it is recommended the field is divided into sections. It can be difficult to make a perfect inspection as is illustrated in Fig. 4.4, which shows a detailed registration in 10 × 10 m grids of three weed species.

International Plant Nutrition Institute (IPNI) (2017, point 15) has some nice thoughts and comments about the requirements and complexity of weed scouting:

> The concept behind scouting for weeds is to provide accurate and timely information needed to make intelligent, cost effective decisions. Moreover, scouting is a key component in the design of effective weed management strategies that help to manage risks by providing information needed to optimize the correct timing of herbicides and accurately monitor weed management successes and failures (Wallace 1994).

> This requires one to carefully think about the dynamic and flexible weed management systems to meet challenging demands. Adaptive sampling strategies (rather than fixed strategies such as grid sampling) are flexible and build on previous information and experience (…) However, we must recognize that here is no single scouting strategy that is best in all situations and that each strategy has advantages and disadvantages (IPNI 2017 point 15).

Table 4.3 Weed scenarios (EPPO codes), their frequency in each of the five crops and required level of control for each of the weed species

Crop	W sc. Area total weeds				Target effecacy (%)
Winter wheat	1	11%		STEME CAPBP MATIN VERPE (basic)	75 50 85 65
	2	21%		Basic + VIOAR	80
	3	32%		Basic + POAAN	85
	4	32%		Basic + APESV	95
	5	5%	100%	Basic + ALOMY	98
	6	25%		GALAPCONAR	85 75
	7	25%		MATIN VIOAR APESV	85 75 85
Spring barley	8	56%		STEME CHEAL GAESS BRSNN LAMSS VERPE (Basic)	80 80 85 80 65 65
	9	33%	89%	Basic + CHEAL	80
	10	22%		AVEFA	95
	11	22%		CIRAR	80
	12	22%		AGRRE	85
Maize for silage	13	56%		STEME CHEAL MATIN POLPE CONAR VERPE (Basic)	85 85 85 85 90 80
	14	28%		Basic + SOLNI	85
	15	28%		Basic + AGRRE	85
	16	14%	125%	Basic + GERdi	80
Winter oilseed rape	17	60%		STEME CAPBP POAAN LAMSS VERPE (Basic)	95 85 90 85 90
	18	10%	70%	Basic + GALAP	95
	19	70%		TRIAE	90
	20	20%		MATIN	95
Sugar beets	21	203%		STEME CHEAL POLPE LAMSS VERPE (Basic)	92 96 96 92 92
	22	34%		Basic + POLAV	96
	23	34%	270%	Basic + SOLNI	96
	24	22%		AGRRE	94

4.3.4 Herbicides

Herbicides may consist of more than one active ingredient, because herbicide products sold to farmers are formulated to optimize their use. The same product may be sold under different trade names and brands in different countries. The CPO-weed calculates recommendations in various units (e.g. litres, kgs, grams and tablets) using local product names. To communicate the CPO recommendations internationally, the product recommendations are transformed to gram active ingredients and the so-called treatment frequency index (TFI). A TFI of 1 is equivalent to one standard dose per ha.

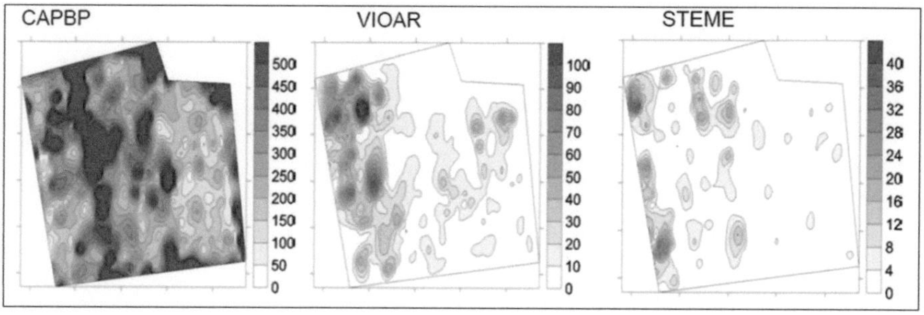

Fig. 4.4 Weed density for six weed species based on 10 m × 10 m grid. CAPBP = *Capsella bursa-pastoris*/Shepherd's Purse, VIOAR = *Viola arvensis*/Field pansy, STEME = *Stellaria media*/Common Starwort (Origin M. Walter/T. Heisel, DJF) (http://pvo.planteinfo.dk/cp/documents/InfoMarkInsp.pdf)

Table 4.4 shows CPO-weed (batch version) recommendations for the 24 weed scenarios.

Table 4.5 shows herbicide use (TFI) and costs (€ha^{-1}) in winter wheat reported by Danish farmers 2015 and calculated with CPO-weed (batch version).

CPO-weed recommendations reduce herbicide use and cost by around 45% and 52% respectively compared to the average herbicide use and estimated costs reported by Danish farmers 2015. It requires, however, 2–3 weed scoutings and planning with a DSS to obtain the potential 33 € ha^{-1} saving.

Some pre-emergence herbicides, like pendimethalin and prosulfocarb, control a broad spectrum of weeds. These herbicides are applied before the actual weed composition in a field can be determined, and selection of the dose will depend on knowledge of previous years' weed infestations. For post-emergence herbicides, on the other hand, more information will be available and generally the potential for reductions in herbicide use is greater. Herbicide reductions are, however, only possible with due consideration of the actual weed flora combined with information on the competitiveness of the crop and climatic factors such as temperature or drought. This information is pivotal because of the very variable susceptibility of weed species to the various herbicides (Sønderskov et al. 2016).

4.4 Weed Detection

4.4.1 Real-Time Weed Scouting

Weed scouting is a prerequisite for using low dose DSS systems, but it takes time and skills to do it manually. Automated, real-time weed scouting and spraying systems that integrate a video devise or camera with weed detection software and algorithms for variable-dose application could solve the scouting and planning problem (Fig. 4.5).

Table 4.4 Herbicide solutions (gram active ingredient per ha) recommended by CPO-weed decision support for five crops and 24 weed scenarios (w). The basic weed compositions are explained in a previous table

w	Crop and weeds	Recommended herbicides and doses (gram per ha)
Winter wheat		
1	Basic	21 g diflufenican +400 g prosulfocarb +0.9 g tribenuron-methyl
2	Basic + VIOAR	21 g diflufenican +400 g prosulfocarb +0.9 g tribenuron-methyl
3	Basic + POAAN	34 g diflufenican +540 g prosulfocarb +0.83 g mesosulfuron +0.27 g iodosulfuron- methyl-natrium
4	Basic + APESV	18 g fenoxaprop-p-ethyl +1.2 g tribenuron-methyl +15 g diflufenican +280 g prosulfocarb
S	Basic + ALOMY	26 g fenoxaprop-p-ethyl +1.4 g tribenuron-methyl +12 g diflufenican +230 g prosulfocarb
6	GALAP CONAR	1.4 g florasulam +2.9 g aminopyralid +52 g 2,4-d
7	MATIN VIOAR APESV	1.2 g metsulfuron-methyl +0.79 g florasulam +2.3 g pyroxsulam +1.3 g sulfosulfuron
Spring barley		
8	Basic	20 g diflufenican +0.65 g tribenuron-methyl
9	Basic + CHEAL	31 g fluroxypyr +0.53 g iodosulfuron-methyl-natrium +0.13 g florasulam +1.4 g diflufenican +0.26 g aminopyralid +4.7 g 2,4-d
10	AVEFA	47 g fenoxaprop-p-ethyl
11	CIRAR	99 g fluroxypyr +2,4 g florasulam
12	AGRRE	840 g glyphosat
Maize for silage		
13	Basic	110 g mesotrion
14	Basic + SOLNI	110 g mesotrion
15	Basic + AGRRE	91 g mesotrion +24 g foramsulfuron +0.81 g iodosulfuron-methyl-natrium
16	Basic + GERdi	370 g bentazon +74 g mesotrion
Winter oilseed rape		
17	Basic	100 g clomazon
18	Basic + GALAP	110 g clomazon
19	TRIAE	24 g propaquizafop
20	MATIN	100 g clopyralid
Sugar beets		
21	Basic	360 g phenmedipham +140 g ethofumesat +96 g desmedipham
22	Basic + POLAV	140 g ethofumesat +140 g phenmedipham +380 g metamitron +96 g desmedipham +5 g triflusulfuron-methyl
23	Basic + SOLNI	260 g ethofumesat +420 g metamitron +84 g phenmedipham +5 g triflusulfuron- methyl
24	AGRRE	87 g propaquizafop

Table 4.5 Herbicide use (TFI) and costs (€ha⁻¹) in winter wheat reported by Danish farmers 2015 and calculated with CPO-weed. HRAC is EPPO herbicide resistance classification

Herbicides		Herbicide use (TFI)		Costs (€ per ha)	
HRAC	Active ingredient	2015	CPO	2015	CPO
A	Clodinafop-propargyl	0.01	0.00	0.5	0.0
A	Fenoxaprop-P-ethyl	0.01	0.11	0.4	3.3
B	Tribenuron-methyl	0.07	0.10	0.6	0.8
B	Sulfosulfuron	0.02	0.02	0.5	0.6
B	Iodosulfuron-methyl-natrium	0.10	0.01	3.2	0.3
B	Flupyrsulfuron-methyl	0.04	0.00	0.7	0.0
B	Florasulam	0.29	0.11	6.9	2.6
B	Mesosulfuron	0.10	0.03	3.7	1.0
B	Pyroxsulam	0.14	0.03	2.3	0.5
B	Metsulfuron-methyl	0.07	0.05	0.9	0.6
C3	Ioxynil	0.05	0.00	1.5	0.0
C3	Bromoxynil	0.05	0.00	1.7	0.0
F1	Picolinafen	0.00	0.00	0.0	0.0
F1	Diflufenican	0.25	0.23	3.1	2.9
K1	Pendimethalin	0.05	0.00	4.8	0.0
N	Prosulfocarb	0.28	0.14	23.0	12.0
O	Fluroxypyr	0.13	0.00	2.3	0.0
O	MCPA	0.02	0.00	0.9	0.0
O	2,4-D	0.01	0.01	0.2	0.2
Z	Aminopyralid	0.01	0.01	0.0	0.0
	Total	1.70	0.86	57.1	24.7
	Reduction med CPO		0.83		32.4
	Relative reduction		49%		57%

It has now become possible to scout for weeds with high-speed cameras like H-sensor (AgriCon, Figs. 4.6 and 4.7) and DAT (Dimensions Agri Technologies AS, Figs. 4.5 and 4.8) mounted on the sprayer. According to Dimensions Agri Technologies AS (DAT) their DAT–sensor will reduce the use of herbicides by an average of 50% — by enabling simultaneous detection and spraying of weeds.

The DAT-Sensor Software image analysis software estimates coverage of broad-leaved weeds and cereals in near-ground RGB images. The algorithm identifies weeds by shape, size, colour and texture. Dicot weeds highlighted in red (Fig. 4.5).

4.4.2 *Weed Mapping with a Camera Mounted on an ATV*

As already mentioned, weed detection (monitoring and scouting), planning and spraying are the three dimensions in chemical weed control with variable doses. Weed detection, planning and spraying, however, do not have to take place

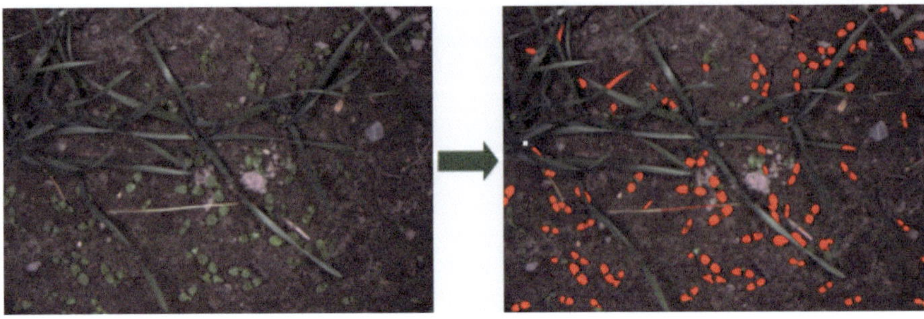

Fig. 4.5 DAT-Sensor Software weed detection (in red: weed identified as dicot weed) (Photos: Dimensions Agri Technologies AS)

Fig. 4.6 H-Sensor weed detection. (in yellow: weed detected as dicot weed) (Photo: AgriCon GmbH)

Fig. 4.7 H-Sensors mounted on a boom sprayer. "Different treatment for grass and dicot weed" (Photo: AgriCon GmbH)

Fig. 4.8 DAT–sensors mounted on a sprayer (Photo: Dimensions Agri Technologies AS)

simultaneously. A weed map produced with automated weed scouting could be a cost efficient alternative to manual weed scouting, and an ideal basis for planning with a DSS like CPO-weed.

The Danish RoboWeedSupport project (Laursen et al. 2017) intends to bridge the gap between the potential herbicide savings using a decision support system like CPO-weed and the required weed monitoring. Their project has examined the cost of performing data collection based on a camera system with three cameras on a 24-m boom mounted on an all-terrain vehicle (ATV) able to drive and record data at up to 50 km h^{-1} with an image quality sufficient for identifying newly emerged grass weeds. Their economic estimates are based on approximately 100 hectares recorded at three different locations in Denmark. With an average image density of 99 images per hectare, the ATV had a capacity of 28 ha per hour, which is estimated to cost 6.6 € ha^{-1}. Alternatively, relying on a boom mounted solution on a tractor, while applying fungicides or fertilizer before the last follow up herbicide application, it was estimated that a cost of 2.4 € per ha is obtainable under equal conditions.

4.5 Conclusion

Site-specific application of pesticides has so far focused mainly on herbicides. The purpose of precision farming technologies in relation to herbicide use is to reduce herbicide cost and environmental impact from spraying, but at the same time to maintain a satisfactory level of weed control. Another purpose is to increase spraying capacity, to reduce the number of sprayer refills and to minimize time spent on weed monitoring. In this chapter, the relevance and profitability of four precision herbicide application technologies, two weed detection technologies and a low dose decision support system (DSS) are analysed. Crop Protection Online-weeds (CPO-Weed) is a decision support system for chemical weed control developed in Denmark and subsequently adjusted to the conditions in several other countries. With lower than recommended herbicide doses provided by CPO-weed, herbicide cost can be reduced by 20–50%. It requires, however, a proper monitoring of weeds, which can be a time-consuming task that, furthermore, requires that the farmer is able to identify the weed species correctly. The current development of high-speed camera and software systems can take over the task of detecting and mapping weeds. However, at the moment no system is available that is able to distinguish and classify all common weed species correctly.

References

Anglund EA, Ayers PD (2003) Field evaluation of response times for a variable rate (pressurebased and injection) liquid chemical applicator. Appl Eng Agric 19:273–282

DST (2016) Economics of agricultural activities 2015. Statistics Denmark (DST), December 2016. http://www.dst.dk/Site/Dst/Udgivelser/GetPubFile.aspx?id=20725&sid=oko2017

Franco C, Pedersen SM, Papaharalampos H, Ørum JE (2017) The value of precision for image-based decision support in weed management. Precis Agric 18(3):366–382

Grisso R, Alley M, Thomason W, Holshouser D, Roberson GT (2011) Precision farming tools: variablerate application. Resource document: Virginia Cooperative Extention. https://pubs.ext.vt.edu/442/442505/442505_PDF.pdf

IPNI (2017) Site-specific management guideline. International Plant Nutrition Institute, USA (IPNI). http://www.ipni.net/ssmg

Jørgensen LN, Noe E, Langvad AM, Jensen J, Ørum JE, Rydahl P (2007) Decision support systems: barriers and farmers' need for support. Bull OEPP/EPPO Bull 37(2):374–377

Laursen MS, Jørgensen RN, Dyrmann M, Poulsen R (2017) RoboWeedSupport: sub millimeter weed image acquisition in cereal crops with speeds up till 50 Km/h. Paper accepted for 19th international conference on precision agriculture. Kyoto, Japan April 27–28, 2017

Lund I, Søgaard HT, Graglia E (2006) Microspraying with one drop per weed plant. Proceedings of Plantekongres 2006, Arhus, Denmark, January 10–11, 2006

Lund I, Christensen S, Jensen LA, Jensen PK, Olsen HJ, Søgaard HT et al (2008) Cellesprøjtning af ukrudt i majs. Pesticide research document no. 123 2008. Ministry of Environment of Denmark. http://www2.mst.dk/udgiv/publikationer/2008/9788770528511/9788770528511/pdf/9788770528528.pdf

Mathiassen SK, Lund I, Kudsk P (2016) Adjuvants for single droplet application of glyphosate proceedings ISAA 2016. https://www.researchgate.net/publication/304497893_ADJUVANTS_FOR_SINGLE_DROPLET_APPLICATION_OF_GLYPHOSATE

Mertz (2017) Personal communication. Leif Trane, Johannes Mertz A/S, Denmark. leif.trane@mertz.dk. +45 2343 5259

MST (2017) Bekæmpelsesmiddelstatistik 2015. Behandlingshyppighed og pesticidbelastning, baseret på salgsstatistik og sprøjtejournaldata. Orientering fra Miljøstyrelsen (MST) nr. 17. Januar 2017. http://www2.mst.dk/Udgiv/publikationer/2017/01/978-87-93529-63-2.pdf

Pedersen SM (2003) Precision farming—technology assessment of site specific input application in cereals. DTU: IPL

Pedersen SM, Fountas S, Blackmore S, Gylling M, Pedersen JL (2004) Adoption and perspectives of precision farming in Denmark. Acta Agric Scand B Soil Plant Sci 54:2–6

Petersen PH, Nørremark M, Jensen PK, Thierry AM, Hørfarter R, Jensen JE, Elbæk J, Lyngvig HS (2017) Udkast til notat om intelligent sprøjteudstyr/præcisionssprøjtning i jordbruget, der kan medvirke til minimering af brugen af pesticider. Memorandum draft on the potential of intelligent spraying equipment and precision spraying to minimize pesticide use in Danish agriculture (Draft January 31st 2017). SEGES P/S and Aarhus University (in prep for a final MST report)

Sønderskov M, Rydahl P, Bøjer OM, Jensen JE, Kudsk P (2016) Chapter 13: Crop protection online—weeds: a case study for agricultural decision support systems. In: Papathanasiou J et al (eds) RealWorld decision support systems, page 303320. Integrated series in information systems 37. Springer International Publishing, Switzerland 2016. https://doi.org/10.1007/9783319439167_13. http://link.springer.com/chapter/10.1007%2F9783319439167_13

van de Zande JC, Achten VT, Kempenaar C, Michielsen JM, van der Schans D, de Boer J et al (2009) SensiSpray: sitespecific precise dosing of pesticides by online sensing. In: van Henten EJ, Goense D, Lokhorst C (eds) Precision agriculture'09. Wageningen Academic Publishers, Wageningen, pp 785–792

Wallace A (1994) High-precision agriculture is an excellent tool for conservation of natural resources. Commun Soil Sci Plant Anal 25:45–49

Walter AM, Heisel T (2001) Precision application of herbicides using injection sprayer systems. In: Grenier G, Blackmore S (eds) Proceedings of the 3rd European conference on precision agriculture. Agro, Montpellier, pp 611–616

Chapter 5
The Economics and Perspectives of Site Specific Irrigation Management in Australia

Robert Farquharson and Jon Welsh

Abstract Automated furrow irrigation is a new technology being developed commercially and offered to farmers in Australia. Improvements in water, fertiliser and labour efficiencies are possible with the more precise management of irrigation water, albeit with initial capital and ongoing management costs. The systematic quantification of potential benefits is a strength of the analysis reported here, which provides information for cotton growers in the Namoi Valley of northern New South Wales (NSW), Australia. Further, an economic investment analysis which considers the benefits and costs over a 20-year period shows potentially favourable returns on investment in this technology.

Not only are there 'private' potential economic benefits to cotton growers, but the estimated reduction in greenhouse gas emissions due to improved efficiency of water and nutrient management can provide a wider 'public' benefit. When this was included in the economic metric, the investment returns are even higher.

This analysis has considered the economics of adopting an improved precision irrigation technology from two different perspectives. Unfortunately the current environmental policy settings in Australia do not allow the wider benefits to be recognised or rewarded. Nevertheless, the environmental gains from improved production efficiency can still be achieved through private decisions.

Keywords Cotton • Automated furrow irrigation • Site specific • Economic • Investment decision • Australia

R. Farquharson (✉)
The University of Melbourne, Melbourne, VIC, Australia
e-mail: bob.farquharson@unimelb.edu.au

J. Welsh
Cotton Research and Development Corporation, Narrabri, Australia

S.M. Pedersen, K.M. Lind (eds.), *Precision Agriculture: Technology and Economic Perspectives*, Progress in Precision Agriculture,
https://doi.org/10.1007/978-3-319-68715-5_5

5.1 Introduction

Precision Agriculture (PA) is a market opportunity that has been substantially addressed by commercial firms developing products for sale to agricultural industries and farmers. Agribusiness firms are leading the emerging market for PA technologies, which are embedded, or embodied, in new products. The PA products are examples of embodied technological change.

These commercial products are available to farmers who make their own private decisions about whether to buy (and adopt the technology). Farmers are well informed about their own particular circumstances and needs, so there seems to be little case for a public evaluation of PA in terms of agricultural benefits, costs and likely adoption. Agribusiness firms also know their markets, target audiences, the relevant agricultural systems and the needs of their customers, so that their commercial research and development activities and investments are made with full knowledge of the potential benefits and risks.

Potentially beneficial aspects of PA management can be identified. One involves improving the precision of application of a variable-rate input such as fertiliser. Several analyses have considered farm profit improvements from single versus variable fertiliser rate PA technologies. These include Paz et al. (1999), Thrikawala et al. (1999) and Babcock and Pautsch (1998), and the returns were found to be only modest in some or most cases. These results are consistent with Pannell's (2006) observation about flat payoff functions from diminishing-returns responses such as fertiliser applied to crops, with implications for economic benefits from PA technologies. Sensor-based N rate calculators have been developed by US Land Grant Universities to predict yields mid-season and make fertiliser recommendations based on predicted crop yields and prices (Prof. H. Zhang, Oklahoma State University, personal communication).

A second aspect of agricultural production systems is the pervading variability in agricultural responses to management and inputs. Any technology that diminishes the uncertainty of responses to management decisions is likely to have potential value, but the payoffs compared to the non-adoption alternative may not be obvious. Agricultural production systems are also characterised by multiple inputs and outputs (joint products), which can complicate the context for a PA technology.

A third issue for agricultural management and PA technologies is the potential to reduce the labour input. Management of dairy cattle for daily milking is a very time-consuming process. Fruit and vine crop pruning and harvest can be automated but there are many orchards and vineyards where manual picking is still used for quality control. Management of irrigation for intensive cotton production (furrow irrigation) can involve labour at all hours of the day or night. Any PA technology that can improve the labour productivity of intensive agricultural management is potentially valuable.

Another aspect of agricultural management is the environmental effects of management practices, and if PA can potentially improve environmental outcomes then there may be considerable 'public' value from new PA technologies. Productivity gains from PA associated with improved precision, addressing uncertainty and enhancing labour productivity can be assessed from a private (business) perspective. If there are also environmental improvements from PA, then they can be assessed from a public perspective.

5.2 Irrigation and Cotton Growing in Australia

Water is a valuable commodity in Australian agriculture given the diverse and substantial agricultural industries, relatively dry climate and extensive land use. Water supplies for Australian agriculture are from precipitation and irrigation. Uncertainty about the availability of irrigation water is widely accepted as the most limiting factor in Australian cotton production systems (Roth Rural 2014). A map of the cotton growing regions of Australia is in Fig. 5.1.

As global and domestic demand for food and other agricultural products increases, there is pressure for cotton growers and irrigation professionals to understand and manage irrigation methods and distribution systems with greater certainty. They not only have to consider potential development of new irrigable land but also re-evaluate technology to manage existing irrigation systems better for cotton pro-

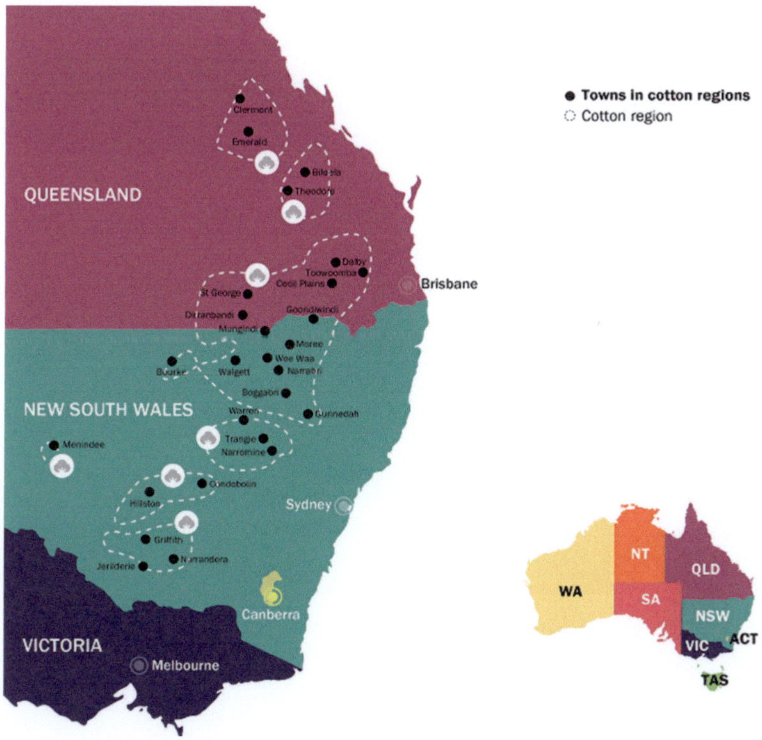

Fig. 5.1 Cotton growing regions of Australia

duction. Because of industry sustainability imperatives requiring more careful use of fossil fuel-derived inputs such as N fertiliser and direct energy, automated furrow irrigation is a potentially desirable technology.

5.2.1 The Importance of Timeliness

Managing irrigation scheduling is critical to resource use efficiency in a water-limited environment. Water for plant growth is provided by rainfall and or irrigation, but for each the amount of water supplied is rarely exactly what is required by the plant. Both the timing and quantity of water applications can vary substantially. Cotton is a summer crop in Australia grown mainly in the 400–800 mm summer rainfall zone. Cotton crops can receive significant amounts of their water needs from rainfall during the growing season. But surface irrigation is vital for profitable cotton production, and furrow irrigation remains the preferred method for irrigating cotton.

Existing furrow irrigation generally involves manual pulling of syphons to initiate water flow (Fig. 5.2) and later inspection of soil wetting down the field to close off water flow. Watering may be conducted during the day or night, and the inspection and close off process may be delayed, leading to over watering (soil profile saturation) and inefficient water use (overflow). Cotton furrow irrigators spend a substantial amount of time completing irrigation tasks. The difficulties associated with irrigation scheduling mean that it is often unrealistic to terminate watering manually, especially at night when it is difficult to check the progress of irrigation water. Automated furrow irrigation (telemetry and automation) includes smart wireless sensors to monitor field conditions and connection to automated gates controlling water flow (Fig. 5.2).

Fig. 5.2 Traditional and automated furrow irrigation

5.2.2 Irrigation Application Efficiency

Furrow irrigation has low capital costs, and is easy to operate and simple to administer. However, because of the manual methods employed there is often a low application efficiency (AE). The AE is defined as the percentage of total water applied that is added to the root zone storage and can be used by the crop (Australian Cotton Industry Development & Delivery Team 2016). Cotton growers are unable to estimate the duration of irrigation accurately before or during the irrigation.

Analysis of a survey of Australian cotton grower practices (Cotton Consultants Australia 2015) found that visual crop monitoring (84%) and weather forecasts (75%) were the main tools for irrigation planning and scheduling. Analysis by Gillies (2012) of 542 surface irrigation evaluations found the average AE to be 75%.

5.2.3 Labour Productivity and Costs in the Cotton System

Historically, Australian agricultural production and productivity have risen in response to pressure from adverse movements in prices received and paid (Nossal et al. (2009)). Adoption of new technologies (new genotypes, changes in land management, increased resource-use efficiency) has been used by Australian farmers to offset the declining terms of trade. This has enabled increases in land and labour productivity, and improved efficiencies through increasing the scale of farm operations (ABARES (2016)). Analysis by Sheng et al. (2016) found that Australian broadacre farms have been able to respond to changing prices and technologies through input substitution (capital for labour) to improve income.

A rapid expansion in Information and Communication technologies has also helped improve farm productivity by increasing output in broadacre agriculture (Salim et al. (2016)). However, while these technologies can reduce costs through optimal use of inputs and improved market participation, labour remains a critical component in the irrigated cotton system. In a recent submission to the Department of Agriculture on behalf of the cotton industry, Vicary (2016) stated that up to two thirds of the industry's 10,000 casual employees are made up of working holiday makers or 'backpackers'. These overseas workers tend to fill labour gaps, particularly in peak times when labour is difficult to access for planting, irrigating and harvesting.

Despite substantial technological advances in cotton harvesting (all-in-one round module picking) displacing a large component of casual labour, cotton production costs have increased over time. Data collated from Boyce and Co. (2016) and Powell (2016) show that casual labour costs ($ ha^{-1}) have increased by 5% year^{-1} over a 20-year period, and more recently by 15% per annum.

5.2.4 *Automated Irrigation Scheduling*

Recent research suggests wireless sensor technologies can be combined with automation to support water use and labour efficiency (Khriji et al. (2014)). An Australian study by Ooi et al. (2010) examined the water savings in an automated irrigated apple orchard using real-time feedback control. There was a 73% improvement in water use efficiency in a pressurised system compared to a baseline.

Koech et al. (2014) investigated methods for real-time control of surface furrow irrigation. This optimisation system estimated the soil infiltration characteristics to control flow meters and water cut-off through a telemetry network. Unlike other approaches to the calibration of soil moisture through remote sensing and a network of probes, the optimisation strategy used a simulation model to estimate infiltration. The simple model assisted the user to optimise all combinations of inflow rate and cut-off time. The main advantage of the infiltration simulation approach to real-time monitoring is the ability to adapt and modify management strategies in real time to suit the prevailing soil conditions and water inflow rate during an irrigation event. The study found water savings using the infiltration simulation technology from the baseline scenario usually used by the farmer, although exact quantities were not calculated.

There appears to be a general acceptance that gains in water use efficiency and labour savings can be attributed to automation and irrigation scheduling technologies (Uddin et al. (2015), Foley (2016)). While the methods for achieving these gains are frequently explored, the quantification of water savings is less clear for broadacre farming, but see Lucas (2016) for sugar cane. Limited research exists on the explicit study of automated irrigation and the connectivity with nutrient use efficiency in agriculture.

5.3 Cotton Nutrition and Managing Nitrous Oxide Emissions

Australian irrigated cotton lint yields are the highest of any major cotton producing country in the world, being about three times the world average. While these yield gains are attributed to improved cotton variety technology (principally for insect resistance), the plant nutrient requirements have also increased. Nitrogen is the most important nutrient in cotton production; it has more effect on yield, maturity and lint quality than any other plant nutrient Hons et al. (2004).

The relationship between irrigation scheduling and N management is finely balanced. Not only can overwatering lead to denitrification of applied fertiliser and the production of nitrous oxide (N_2O, a potent greenhouse gas (GHG)), but cotton lint yield penalties during soil saturation can be 12 kg lint/ha/hour.

The interaction between irrigation scheduling, cotton yield and Nitrogen Use Efficiency (NUE) has gained considerable attention from policy makers and the cotton industry. Improvements in NUE in agricultural production are seen as critical for addressing the triple challenges of food security, environmental degradation and climate change (Zhang et al. (2015)). Optimising NUE in a system to go beyond traditional yield barriers is challenging for policy makers, industry and cotton growers.

The N_2O from denitrification increases exponentially as fertiliser rates increase, making up to 3.5% of these losses when N rates are applied between 280 and 320 kg N per hectare. A recent survey by Cotton Consultants Australia (2015) of cotton grower practices showed the N application rate exceeded 351 kg ha^{-1} for approximately half the surveyed planted area. Baird (2016) showed how irrigation management strategies influence the N cycle within the irrigated cotton system. By regulating the amount of water applied during an irrigation event and optimising N applications resulted in a more efficient N uptake and reduced system losses of N, enhancing both yield and crop gross margin. Losses from the system ranged between 29 and 47 kg N ha^{-1} and occurred primarily from the first irrigation. Hence it is important to quantify possible mitigation of GHGs through irrigated furrow automation.

During the last decade, many Government- and industry-funded agricultural research and extension initiatives have been specifically aimed at improving NUE on cotton farms. These include the 'Nitrous Oxide Research Program', 'Carbon Farming Initiative Extension and Outreach', 'Action on the Ground'and 'Filling the Research Gap'. A study on the future of Australian agricultural productivity by Grundy et al. (2016) suggested that climate change mitigation policy settings may have a strong impact over the period to 2050.

5.3.1 *Perspectives on Improved Irrigation Efficiency in Australia*

From the above discussion there are potential advantages for Australian cotton growers from implementing improved (automated) furrow irigation. Such improvements can include less water loss (overflow from the tail of the field), improved crop yield (less soil profile saturation), improved NUE (less denitrification) and improved labour efficiency. Furthermore, these improvements can be assessed economically (in an investment or cost-benefit analysis (Sinden and Thampapillai (1995)) to determine whether a private (cotton grower) benefit is likely from adoption of this PA technology. The efficiency improvements (AE and NUE) comprise the main production benefits for a private (investment) analysis. It can be difficult for an individual cotton grower to assess the combined improvements in water use, N use and labour efficiency when considering an investment in improved irrigation technology.

But the potential NUE improvements which reduce GHG emissions also provide a wider social or public benefit from adopting the technology. If the reduced GHG emissions can be 'priced' and added to the private benefit calculation, then the private investment analysis can be extended to a public benefit assessment.

5.4 Method of Analysis

The private benefit evaluation of automated furrow irrigation used conventional investment analysis for implementation of this PA technology for a typical or representative cotton farm in the lower Namoi River region of northern NSW, Australia. A representative farm analysis can develop information for use by a similar group of cotton growers in the region.

5.4.1 *Investment Analysis*

A discounted cash flow (DCF) analysis (Sinden and Thampapillai 1995) was conducted for an investment in automated furrow irrigation for the representative farm, upgrading from traditional furrow irrigation. The DCF measures of net present value (NPV) and internal rate of return (IRR) were estimated. The analysis was conducted on a $ ha^{-1} basis (i.e. at the farm enterprise (field) level) since cotton production is the major production enterprise on cotton farms. It focused on cost savings associated with the automated furrow irrigation investment. The farm-level efficiencies associated with the investment were calculated over a 20-year period. The physical (capital) requirements were for an extra irrigation channel, within-bank pipes, automated irrigation gates, in-field sensors, and telemetry and computer software programs.

There were benefits and costs associated with on-farm investment in automated furrow irrigation. The benefits categories included the avoidance of labour costs for cotton irrigation, of labour costs for semi-irrigated wheat, of costs of excess water supply and of costs of excessive fertiliser use. Crop yields were assumed to be unchanged. The cost categories included installed automation (capital costs), annual repairs (solenoids, channel maintenance), and opportunity costs of lost land for the additional channel. Projected increases in labour costs over time were included based on recent experience in labour cost increases. A discount rate of 7% (NSW Premier and Cabinet 2016) was used, with sensitivity analysis of 4 and 10%. Private discount rates may be higher than public (government funded investment) rates. A 7% private and 4% public discount rate are most suited for this analysis, but the sensitivity tests show results for all three rates.

5.4.2 *Characteristics of the Lower Namoi Representative Farm*

A representative farm identified by Powell and Scott (2015) was used as a basis for the analysis. The representative farm included information gathered from available data, local consensus groups and assumptions regarding the size of a typical farm and other resources, such as labour, overhead costs, assets and liabilities and the nature of the cropping rotation used. The breakdown of land use and water resources is shown in Table 5.1. The farm labour supply is also shown in Table 5.1 for the whole year. Casual labour inputs were included in the crop gross margin budgets which underlie the representative farm model.

The investment analysis was conducted for a field within this representative farm, i.e. on a per hectare (ha) basis. The crop sequence is cotton-wheat-long fallow, as shown in Fig. 5.3. Cotton is planted during the spring season in October in the first year. Soon after cotton picking in April the field is sown to wheat. After wheat harvest in the following November the field is fallowed for 10 months to accumulate stored soil moisture from rainfall before being returned to cotton. It was assumed that this crop rotation continues for 20 years and that the crop yields do not change over that period.

Table 5.1 Characteristics of the Lower Namoi representative farm

Farm Area	Metric	Size
Total farm area	ha	1203
Irrigable land	ha	782
Minimum area irrigated annually	ha	250
Planned automated irrigation area	ha	500
Area farmed – dryland	ha	180
Area grazed	ha	120
Water resources		
Groundwater allocation	ML	2500
Namoi River allocation	ML	1600
Water storage capacity	ML	900
Whole farm annual labour		
Owner manager	No of weeks	50
Permanent employee	No. of weeks	48
Casual labour	Factored into crop gross margins	

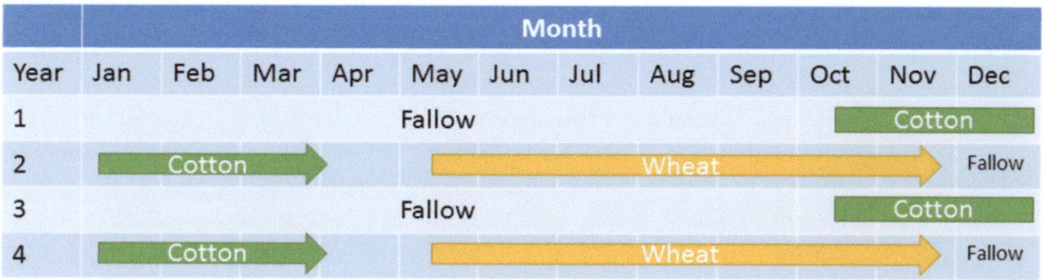

Fig. 5.3 Schematic diagram of the first 4 years of a 20-year continuous irrigated cropping rotation on the representative farm

5.4.3 Modelling Hydrology and Greenhouse Gas Effects

5.4.3.1 Hydrology

'Howleaky' (McClymont et al. (2008)) is a modelling tool for analysis of the impact of different land uses and management for changed water balance, deep drainage, soil erosion and water quality outcomes (Melland et al. (2010), Carroll et al. (2012)). Water use in irrigated agriculture is generally measured in mega litres (ML).

The Howleaky decision support system was run using irrigation and agronomic parameters consistent with known cotton industry practices and calibrated with historical climate data for Wee Waa (NSW) from the Australian Bureau of Meteorology SILO (2016) database. A 20-year irrigation (1996–2015) simulation using Howleaky estimated water savings from business as-usual (manual siphon irrigation) compared with automated delivery and shut-off from the new technology. Research by Roth Rural (2014) and Gillies (2012) underpins the chosen industry irrigation practices and application efficiency assumptions. Cotton N fertiliser application rates for the baseline scenario were derived from a case study farm in the Lower Namoi and recent industry survey data (Cotton Consultants Australia 2015). Although semi-irrigated wheat was included in the cropping system, irrigated cotton was the focus of water savings in the water balance model. Table 5.2 summarises soil characteristics and key input parameters for the irrigation and N components of the study. The optimal soil moisture deficit was chosen from a study by Baird (2016).

Cotton crop modelling was incorporated into the Howleaky model through use of the Agricultural Production Systems Simulator (APSIM) modified from traditional cereal-based characteristics to simulate growth characteristics of the cotton crop (McClymont et al. (2008). Estimates of vegetative parameters were used to determine transpiration, soil evaporation and infiltration.

5.4.3.2 Greenhouse Gases

Greenhouse gas emissions scenarios were estimated using the FarmCarbon Calculator. This enables the user to investigate GHG abatement options through modifications of farming input and yield assumptions and 'what if' scenarios for a

Table 5.2 Soil characteristics and input parameters for applied N fertiliser and water balance model

Scenario	Soil – plant available water holding capacity	Soil moisture deficit irrigation trigger	Irrigation application	Applied N Fertiliser rate kg/ha
Baseline	200 mm – Grey vertosol[a]	70 mm	Field capacity	300
Automated		70 mm	70 mm	250

[a]A vertosol is a soil with a high content of expansive clay, the alternative shrinking and swelling of which causes self-mulching. Vertosols typically form from basic rocks such as basalt (The Australian Soil Classification, CSIRO)

Table 5.3 Assumptions for the economic investment analysis

Avoided labour costs of automation, calculated for the 2-year cotton-wheat-long fallow rotation, made up the majority of the project benefits (\$290/ha), followed by water savings (\$60/ha) and avoided fertiliser costs (\$50/ha)
The installed automation investment cost was \$750/ha in the first year. Other operating costs include maintenance of 1%/year (\$8/ha) and an opportunity cost of reduced farmland area from an extra delivery channel calculated at 0.6% per hectare of land. Foregone profits from farming have been derived from annual irrigated cotton and semi-irrigated wheat gross margins for the lost land area, amounting to \$12/ha/year
Future costs and benefits were increased by 2.5%/year
Avoided labour costs were increased by 5%/year. This is consistent with analysis derived from the Boyce and Co. (2016) cotton comparative analysis for employee wages over the last 20 years
A carbon price was included in the public economic analysis. A price of \$25/t CO2e was used for the analysis, which was the original level at the commencement of the carbon price policy. However, \$10/t CO2e is the current auction price from the latest Emissions reduction fund
The cost of water varies considerably depending on the source, location and number of times water is pumped. A cost of \$60.30/ML was used assuming pumping from the regulated Namoi River (fees \$32.24/ML) with water 'lifted' twice using diesel pumps (pumping cost \$28.06/ML). This pumping cost takes into consideration fuel and maintenance. Pumping costs can be in excess of \$120/ML for bores.

Note: *ML* Megalitres, *\$* Australian dollars

range of individual farm enterprises and land use types. The default settings in the FarmCarbon Calculator were based on life cycle assessment (LCA) research undertaken by Visser et al. (2015) and riparian research from Smith et al. (2014). The calculator was used to assess changes in emissions at the farm level under a business-as-usual scenario and with automated furrow irrigated technology. FarmCarbon scenarios were used to measure the changes in soil carbon from applied N fertiliser rates from 300 kg N ha^{-1} reduced to 250 kg N ha^{-1}.

5.4.4 *Assumptions for Investment Analysis*

A summary of model assumptions for the economic investment analysis is in Table 5.3.

5.5 Results

5.5.1 *Water Savings*

The Howleaky water balance model was used to quantify potential water savings from higher irrigation AE for irrigated cotton. Simulations were run for a continuous cotton crop rotating every 2 years between 1995 and 2015 for a range of

Table 5.4 Results of Howleaky water balance scenarios from a range of irrigation cotton scheduling practices

Scenario (deficit irrigation trigger)	Cotton crop transpiration (mm)	Field evaporation (mm)	Irrigation applied (ML/ha)	Run-off (ML/ha)	Deep drainage (ML/ha)	Water savings from 70 to 70 mm (ML/ha)
70–70 mm	776	391	6.33	0.34	0.25	0
70 mm-FC	850	356	7.04	0.35	0.54	1.00
70 mm-Sat	865	354	9.29	0.51	2.48	6.44

conditions including: (i) 70 mm deficit irrigation trigger to 70 mm application, (ii) 70 mm deficit irrigation trigger to field capacity (FC) (FAO 2016), and (iii) 70 mm deficit irrigation trigger to soil saturation. These scenarios are shown in Table 5.4. The first irrigation scenario (i) depicts the most accurate irrigation practice, i.e. the amount of water required to bring the soil deficit to zero is applied. Irrigation to field capacity in scenario (ii) is when the drainage has stopped and the large soil pores are filled with both air and water while the smaller pores are still full of water. Scenario (iii) shows the application of irrigation water until in all soil pores are filled with water. In this case the soil is saturated. Each scenario has a different water balance scenario of water use and water losses. The water balance scenario was not calculated for the semi-irrigated wheat crop in the time series rotation because there was only one small watering per wheat crop. Results from the water balance modelling are summarised in Table 5.4.

The results of the irrigated cotton simulations indicated a water saving of 1 ML per hectare from the baseline scenario of irrigating to field capacity. The majority of savings occurred from reductions in applied irrigation water (0.71 ML ha^{-1}) when compared with savings from deep drainage (0.29 ML ha^{-1}). Interestingly, the third scenario of measured water savings from 70 mm deficit applied to saturation showed a five-fold increase in deep drainage losses and water saving of 6.44 ML ha^{-1} from the baseline scenario (70 mm–70 mm). These results highlight the notion that improving scheduling practices from low AEs can lead to significant water savings. To quantify these potential savings; irrigated cotton gross margin analysis by Powell (2016) found the variable cost of irrigation water to equal $60.30 per ML ha^{-1}.

5.5.2 *Changes in GHG Emissions*

The change to automated furrow irrigation enables more accurate irrigation application through shut-off capabilities compared to manual pulling of siphons. Automation reduces opportunities for saturation from overwatering and in-crop rainfall events, the production of N_2O emissions, nitrate leaching and nitrate run-off in tail water (MacDonald et al. (2015) MacDonald et al. (2016)) as well as reduced yield losses from overwatering. The reduced rate of applied N fertiliser has been chosen as the

Table 5.5 Changes in GHG emissions on a lower Namoi Valley representative farm, reduction in applied N fertiliser under automated furrow irrigation

Scenario		Kg/CO_2e/ha/year			
	Fertiliser applied kg N/ha	Nitrate deep drainage	Direct N_2O emissions from cotton	Total crop rotation effect	Change in emissions from baseline
Baseline	300	58.2	4170	4963	0
Automated irrigation	250	51.3	2780	4056	−907

economic optimum from research undertaken by (Baird 2016). A decision support program (Visser 2016) was used to calculate changes in kg CO_2e ha^{-1} year^{-1} under the baseline and automated irrigation scenarios. The estimated changes in GHG emissions from automated furrow irrigation are shown in Table 5.5.

Over the two-year cropping cycle, the automated irrigation scenario resulted in lower GHG emissions (4056 kg ha^{-1} year^{-1}) compared to the reduced rate of fertiliser under the baseline scenario (4963 kg ha^{-1} year^{-1}) due predominantly to the high global warming potential of N_2O emissions. The total N_2O emissions were reduced substantially due to the exponential nature of the emissions factor curve used in the analysis (Visser et al. 2015).

5.5.3 *Economic Analysis*

Results of the investment analyses are presented in Table 5.6. For a private cotton grower the investment in automated furrow irrigation over 20 years is very positive with an IRR of 29% and NPVs of \$2965, \$1979 and \$1334 ha^{-1} at 4, 7 and 10% discount rates, respectively.

When the GHG emissions reductions were included with a carbon price of \$25 t^{-1} CO_2e the 'social' benefit was higher with an IRR of 32% and NPVs of \$3262, \$2203 and \$1510 ha^{-1} at the different discount rates. At a lower carbon price of \$10 t^{-1} CO_2e the social benefit was still very positive with an IRR of 31% and NPVs of \$3083, \$2069 and \$1404 ha^{-1} at the different discount rates.

5.6 Discussion

The key task of farm management is making choices between alternatives. For irrigated cotton growers balancing productivity, profit and environmental responsibility, there is a relationship between water AE, NUE, farm labour and capital inputs. Within the farm management team, numbers, ages and skills of farming families vary, as do the methods used to meet peak workloads and the skills available to

Table 5.6 Results of investment analyses

Item	Unit	Private investment			Public investment ($25/t CO_2e)		
Discount rate	%	4	7	10	4	7	10
NPV	$/ha	2965	1979	1334	3262	2203	1510
IRR	%	29	29	29	32	32	32
					Public investment ($10/t CO_2e)		
Discount rate					4	7	10
NPV					3083	2069	1404
IRR					31	31	31

conduct specialised tasks (Malcolm et al. (2005)). Skills in successful operation of agricultural technology are critical for irrigation management of high input and highly mechanised crops such as cotton. Historical data show that the cost of employee wages has increased at double the rate of inflation since 1997, a large portion of which can be attributed to furrow irrigation tasks.

Furrow irrigation remains the preferred method for irrigation of cotton in Australia due to the low capital cost, ease of operation and simplicity. However, manual methods often result in low AEs because of the inability of farmers to estimate the required irrigation duration accurately before or during the irrigation. In a review of water efficiency and productivity of irrigated cotton, Roth et al. (2013) found 80% of Australia's cotton-growing area use irrigated gravity surface-irrigated systems. Over the last decade there has been increased interest in bankless-channel, drip irrigation and lateral move machine irrigation systems. While bankless-channel systems provide an alternative to siphons and offer immediate labour savings, the initial installation costs can outweigh benefits. Pressurised systems have shown improved water use efficiency compared to furrow irrigation owing to water savings from increased ability to capture rainfall in-season and less in-field deep drainage below the root zone.

The high capital and energy costs associated with pumping remain constraints to both drip and overhead irrigation systems. Cotton industry extension staff acknowledge that changing irrigation systems involves a major decision in an environment of water-allocation uncertainty, cost of system upgrade and higher energy costs from pumping. Irrigation AE and NUE are inextricably linked; industry research findings on NUE also show potential for improvement. The relationship between water management and fertiliser management affects both crop returns and losses to the environment in the form of leaching, removal of nitrates from the field and N_2O emissions.

This study has presented an integrated approach to assess the economic and environmental benefits and costs of installing automation in a furrow irrigated cotton system enabling evaluation of GHG mitigation and farm management responses. This method was applied to a representative cotton farm in the lower Namoi Valley of NSW to assess the trade-offs between labour, fertiliser and water use when installing an innovative irrigation technology. Using FarmCarbon GHG modelling, results showed almost 1 tonne ha^{-1} $year^{-1}$ of CO_2e abatement was obtainable through optimising N fertiliser rates, recommended practices and irrigation deficits without being yield limiting (Baird (2016), Visser et al. (2015)). However, improvements in NUE and irrigation AE are not the only advantages of automated furrow irrigated. In addition, operator error from manual labour may be reduced resulting in greater accuracy and timing. The innovative technology examined in this study may facilitate dual environmental and economic outcomes where supply of labour is limited. On this type of farm, labour is the main limitation for furrow irrigation management.

The results of the economic assessment indicate adoption of automated technology offers net benefits and good investment returns. The investment analysis showed that avoided labour costs are the most important aspect of project returns, followed by water savings.

Avenues for further research include the application of smaller scale (< 500 ha) automated irrigation scheduling and analysis of the optimal row length of a field to assist in reducing installation cost and improved water application. A larger data set of labour savings across a range of farm sizes and cropping rotations would also enhance the rigour of the proposed investment. Optimising N and water use combinations in a number of different climates and soil types is also possible.

5.7 Conclusions

Although cotton growers are generally innovative and attuned to financial imperatives in their farm investment and management decisions, a technology such as automated furrow irrigation has a number of beneficial aspects which may be difficult to combine in an investment analysis. These include water savings, fertiliser efficiency, labour savings and environmental improvements.

In this paper we have included these various aspects in an investment analysis. An analysis such as this, which systematically assesses these potential benefits and combines them into the metric of an economic cost-benefit analysis, can be valuable to cotton growers in making investment decisions. In a private benefit context, the analysis presented here has shown potentially strong returns from such an investment.

The enhanced efficiency associated with automated furrow irrigation can also reduce GHG emissions, and when these were included by valuing them with a typical carbon price the investment was even more appealing. However, the current policy settings in Australia do not provide for individual cotton growers to gain credit for such investment. Hence the 'social' investment returns remain an interesting but hypothetical result.

References

ABARES (2016) Australian Commorities, Australian Government Department of Agriculture and Water Resources, Canberra

Australian Cotton Industry Development, Delivery Team (2016) Waterpak: a guide for irrigarion management in cotton and grain farming systems. In: Cotton Research and Development

Corporation (ed) Online. http://www.cottoninfo.com.au/sites/default/files/documents/WATERpak.pdf
Babcock BA, Pautsch GR (1998) Moving from uniform to variable fertiliser rates on iowa corn: effects of rates and returns. J Agric Resour Econ 23:385–400
Baird J (2016) Identifying practical solutions to optimise NUE and WUE in cotton production, Filling the Research Gap 1601, Final Report prepared for the Department of Agriculture
Boyce and Co (2016) Australian cotton comparative analysis (1997–2016), Online. http://www.boyceca.com/boyce-newsfeed/category/australian-cotton-comparative-analysis
Carroll C, Waters D, Vardy S, Silburn DM, Attard S, Thorburn PJ, Davis AM, Halpin N, Schmidt M, Wilson B, Clark A (2012) Paddock to reef monitoring and modelling framework for the Great Barrier Reef: Paddock and catchment component. Marine Pollution Bulletin 65(4–9):136–149
Cotton Consultants Australia (2015) Qualitative report on the2014–2015 cotton season: a survey of consultants, Cotton research and development corporation. Online. http://crdc.com.au/publications/qualitative-report-2014-15-cotton-season-survey-consultant
FAO (2016) Irrigation Water Management: Introduction to irrigation, Chapter 2 – Soil and water, Online. http://www.fao.org/docrep/r4082e/r4082e03.htm
Foley J (2016) Switching to automated furrow irrigation, National Centre for Engineering in Agriculture, University of Southern Queensland
Gillies M (2012) Benchmarking furrow irrigation in the Australian cotton water story: a decade of research and development 2002–2012. Cotton Cooperative Research Centre, Narrabri
Grundy MJ, Bryan BA, Nolan M, Battaglia M, Hatfield-Doddos S, Conner JD, Keating BA (2016) Scenarios for Australian agricultural production and land use to 2050. Agric Syst 142:70–83
Hons FM, Lemon RG, Nichols RL, Jahn RL, Stapper JR (2004) Managing Nitrogen fertilization in cotton. Agricultural Communications Texas A&M University, College Station
Khriji S, Kanoun O, El Houssaini D, Vieweger C, Jmal MW, Abid M (2014) Precision irrigation based on wireless sensor network. IET Sci Meas Technol 8:98–106
Koech RK, Smith RJ, Gillies MH (2014) A real-time optimisation system for automation of furrow irrigation. Irrig Sci 32:319–327
Lucas W (2016) Automated furrow irrigation economic case study, Burdekin region, Queensland Department of Agriculture and Fisheries. https://publications.qld.gov.au/dataset
MacDonald BCT, Rochester IJ, Nadelko A (2015) High yielding cotton produced without excessive nitrogen oxide emissions. Agron J 107:1673–1681
MacDonald BCT, Chang YF, Warnecke S (2016) Potential contributions of surface and ground water to nitrous oxide emissions from irrigation cotton production systems. Agric Water Manag 168:78
Malcolm B, Makeham J, Wright V (2005) The farming game: agricultural management and marketing. Cambridge University Press, Port Melbourne
McClymont DJ, Freebairn DM, Rattray DJ, Robinson JB, Silburn DM, White S (2008) Howleaky 2008: exploring water balance and water quality implications of different land uses, Online. http://www.howleaky.net
Melland AR, Vigiak O, Roberts AM, Rattray D, Whitford J (2010) Evaluation of a static water balance model in cropped and grazed systems of temperate Australia. Environ Model Softw 25:1682–1691
Nossal K, Shiji Z, Yu S, Gunasekera D (2009) Productivity movements in Australian agriculture. Aust Commodities 16:206–216
NSW Premier, Cabinet (2016) NSW government program evaluation guidelines. NSW Government, ISBN: 978-0-7313-5499-3, Sydney
Ooi SK, Cooley N, Mareels I, Dunn G, Dassanayake K, Saleem K (2010) Automation of on-farm irrigation: horticultural case study. IFAC Proc Vol 43:256–261
Pannell D (2006) Flat earth economics: the far-reaching consequences of flat payoff functions in economic decisiuon making. Rev Agric Econ 28:553–566
Paz JO, Batchelor WD, Babcock BA, Colvin TS, Logsdon SD, Kaspar TC, Karlen DL (1999) Model-based technique to determine variable rate nitrogen for corn. Agric Syst 61:69–75

Powell J (2016) Cotton gross margins, Online. http://www.dpi.nsw.gov.au/__data/assets/pdf_file/0003/577902/cotton-gross-margins-2014-2015-central-and-northern-nsw.pdf

Powell J, Scott F (2015) Irrigated cotton rotations in the Lower Namoi Valley: a whole-farm economic comparison. Aust Farm Bus Manag J (Under review)

Roth Rural (2014) Cotton growing practices: findings of CRDC's survey of cotton growers, Cotton Research and Development Corporation. http://www.crdc.com.au/sites/default/files/pdf/2014%20Cotton%20Practices%20Survey%20REPORT%20-%20Jan%202016.pdf. Narrabri

Roth G, Harris G, Gillies MH, Montgomery J, Wiggington D (2013) Water-use efficiency and productivity trends in Australian irrigated cotton: a review. Crop Pasture Sci 64:1033–1048

Salim R, Mamun SAK, Hassan K (2016) Role of communication technologies in broadacre agriculture in Australia: an empirical analysis using panel data. Aust J Agric Resour Econ 60:243–264

Sheng Y, Davidson A, Fuglie K, Zhang D (2016) Input substitution, productivity performance and farm size. Aust J Agric Resour Econ 60:327–347

SILO (2016) SILO Climate data. Queensland government, Web page

Sinden JA, Thampapillai DJ (1995) Introduction to benefit-cost analysis. Longman, Melbourne

Smith RJ, Visser F, Welsh J, Vogel S, Trindall J (2014) Carbon neutral cotton farms. Australian Cotton Conference, Gold Coast, Australia

Thrikawala S, Weersink A, Kachanoski G, Fox G (1999) Economic feasibility of variable-rate technology for nitrogen on Corn. Am J Agric Econ 81:914–927

Uddin J, Gillies M, Smith R (2015) Smart automated furrow irrigation of cotton. In: Irrigation Australian regional conference, Western Sydney

Vicary C (2016) Cotton Australia submission to working holiday visa review, Sydney

Visser F (2016) FarmCarbon [Online]. Available: http://n-smart.com.au/. Accessed 8 Oct 2016

Visser F, Dargusch P, Smith C, Grace PR (2015) Application of the crop carbon progress calculator in a 'farm to ship' cotton production case study in Australia. J Clean Prod 103:675–684

Zhang X, Davidson EA, Mauzerall DL, Searchinger TD, Dumas P, Shen Y (2015) Managing nitrogen for sustainable development. Nature 528:51–59

Chapter 6
Auto-Steering and Controlled Traffic Farming – Route Planning and Economics

Claus G. Sørensen, Efthymios Rodias, and Dionysis Bochtis

Abstract Agriculture nowadays includes automation systems that contribute significantly to many levels of the food production process. Such systems include GPS based systems like auto-steering and Controlled Traffic Farming (CTF). These systems have led to many innovations in agricultural field area coverage design. Integrating these advancements, two different route planning designs, a traditional and an optimised one, are outlined and explained in this chapter. Four different machinery scenarios were tested in four fields each, and the main aim was to compare the two different route planning systems under economic criteria and identify the best operational route coverage design criterion. The results show that there are significant reductions in operational costs varying from 9 to 20%, depending on the specific machinery and field configurations. Such results show the considerable potential of advanced route planning designs and further optimization measures. They indicate the need for research efforts that quantify the operational and economic benefits by optimising field coverage designs in the headlands, turnings or obstacles avoidance according to the actual configuration to minimize the non-working activities and, as a consequence, the overall operational cost.

Keywords Field coverage • Financial feasibility • Cost • Route planning • Optimisation

C.G. Sørensen (✉) • E. Rodias
Department of Engineering, Faculty of Science and Technology, Aarhus University, Aarhus, Denmark
e-mail: claus.soerensen@eng.au.dk

D. Bochtis
Department of Engineering, Faculty of Science and Technology, Aarhus University, Aarhus, Denmark

Centre for Research and Technology-Hellas (CERTH), Institute for Research and Technology – Thessaly (IRETETH), Volos, Greece
e-mail: dionysis.bochtis@eng.au.dk

S.M. Pedersen, K.M. Lind (eds.), *Precision Agriculture: Technology and Economic Perspectives*, Progress in Precision Agriculture,
https://doi.org/10.1007/978-3-319-68715-5_6

6.1 Introduction

Automation systems in modern agriculture are included in any kind of agricultural machinery and tractors. Many different types of technologies such as radio frequency, laser, machine vision and GPS have been tried in the navigation of agricultural vehicles (Bochtis et al. 2014; Sørensen and Bochtis 2010; Sørensen et al. 2010; van Zuydam and Sonneveld 1994). The GPS-based navigation systems are the only navigation technologies that have become commercially available for navigation of agricultural vehicles. There are two types of GPS-based guidance systems; the GPS guidance-aided systems and the fully automated or 'hands-free' GPS guidance systems that actually steer the tractor with the driver only supervising it. The fully automated system is capable of driving the tractor in a straight line through the field with a lateral accuracy of less than 2 cm. This system uses a very accurate real-time kinematic (RTK) GPS receiver. The RTK-GPS achieve good geopositioning accuracy of a few centimetres. To achieve such accuracy in practice, a GPS base station located close to (10 km) the mobile unit and a radio data link (Gan-Mor and Clark 2001) are required. This system can work with any field and operation, including planting, cultivating and harvest (Batte and Ehsani 2006). The position information from RTK GPS can be used not only for guidance but also for other applications such as seed mapping, controlled traffic, controlled tillage (Chesworth 2008). The RTK-GPS technology systems have been established and used in many different countries throughout Europe (mostly in Northern and Central Europe) over the last 20 years or more (Engfeldt 2005). Auto-guidance field machinery systems in parallel with GPS are used little even in Northern Europe according to recent surveys; it varies from 2 to 24% of the respondents in Finland, Germany and Denmark (Lawson et al. 2011). One of the disadvantages of the use of these technologies is the cost of management and maintenance and, of course, the cost of investment making their use more affordable for large than for small farms (Lawson et al. 2011).

Modern agricultural machinery is equipped with many controls, therefore, operator fatigue is a serious concern. Automatic guidance can reduce operator fatigue and improve machinery performance by reducing overlap or 'skips' during field operations such as tillage and chemical applications (Tillett 1991). With automatic guidance, companies and farmers report that they are able to carry out most field operations in row crops on flat land with greater accuracy than manually steered systems. A typical increase in field capacity is around 15%. Another advantage of the system is particularly noticeable during low-visibility conditions (night time or fog). The present accuracy in row operations can enhance the placing of chemicals in narrow bands or cultivating close to the plant line. Furthermore, use of RTK-GPS guidance to work along contour lines in hilly and rough terrain can reduce erosion and provide additional benefits (Gan-Mor and Clark 2001). Finally, by using autosteering systems, there are many economic and environmental benefits such as lower energy consumption and lower CO_2 emissions (Batte and Ehsani 2006).

Controlled Traffic Farming (CTF) systems are based on the principle that all the traffic inside the field is restricted to specific wheel tracks (tramlines) only. This can be achieved only by using accurate guidance systems i.e. auto-steering control and by aligning the machinery width with the tramline width. Apart from the investment cost of the CTF system, there are many significant benefits. The CTF systems were first introduced because of soil compaction caused by heavy agricultural machinery and tractors. Soil compaction causes reduction of soil infiltrability, conductivity, porosity and aeration and increases bulk density, which implies increased fuel consumption because of the increased pulling force required, which wastes energy (Gan-Mor and Clark 2001).

By using permanent wheel tracks in CTF, all the above problems are avoided because of the specific routes that are determined from the establishment of the crop in the field. In addition, time savings and material savings can reach 10–20% (Kroulík et al. 2011). Additional benefits include increased water retention in the soil, and also the total water runoff from the field is considerably less than in conventional systems. In conclusion, some of the advantages stemming from the implementation of CTF systems are: lower fuel consumption for field operations and cultivation, lower fuel consumption for driving over the soil, better seedbeds, improved soil structure, better fertilizer use efficiency, reduced quantities of agrochemicals, potential to retain more organic matter and living organisms and reduced CO_2 emissions.

6.2 Route Planning Design

Route planning regards the determination of a route that should be followed in the field with minimal costs. In agricultural field operations, the route planning problem is also encountered by operators that have to make a decision on how to traverse the field work tracks to minimize the non-working distance, time and cost. In conventional agriculture, the routes that are followed by agricultural machinery to cover a field area can be implemented several times without being designed properly. The most efficient route planning that should be followed on a given field area should be designed according to many factors such as the lowest fuel consumption, minimization of the non-working distance or non-working time, and as a consequence the minimization of the non-working cost. Route planning can be designed both in conventional and CTF systems given that basic automation systems such as auto-steering systems and GPS navigation exist.

Because of the requirement of creating practices for optimized field coverage, a new pattern has been suggested called B-pattern (Bochtis 2008). The B-patterns are defined as: algorithmically-computed sequences of field work tracks that completely cover an area and that do not follow any pre-determined standard motif, but in contrast are the result of an optimization process under one or more selected criteria (Bochtis et al. 2013). In B-patterns, the best result of the optimisation approach depends on the specific combination of the kinematics and dimensions of the mobile unit, the field shape, the operating width and the optimisation criterion or criteria that will extract the optimal sequences that should be followed. The B-patterns have been tested for an autonomous agricultural vehicle and have shown under the criterion of minimized non-working distance that this distance can be minimized up to 50% for a series of different field operations (Bochtis et al. 2015; Bochtis and Sørensen 2009, 2010 Bochtis and Vougioukas 2008; Bochtis et al. 2009b).

In agricultural operations, there are a number of constraints that must be taken into account such as soil compaction, the fact that a typical agricultural machine cannot usually operate while manoeuvering, and operating while following contour lines.

In addition, the fieldwork pattern followed in previous treatments or by other machinery types is another significant problem regarding route planning. Consequently, area coverage planning is mostly determined by agronomic parameters and constraints. For this reason, the whole problem of area coverage planning in field operations is considered as a sequence of sub-problems:

(a) Field area disintegration i.e. disintegration of the coverage region into sub-fields when needed and generation of headlands in the field or and in the sub-fields.
(b) Determination of the driving direction in each sub-field.
(c) Field track generation. It determines how the set of parallel field tracks is generated.
(d) Route planning over the geometrical representation extracted from the above sub-problems. The resulting route refers to the areal cover of sub-fields, meaning the generation of a track that covers each sub-region ensuring that the vehicle covers the main core of the in an optimum way, according to an optimization criterion (i.e. the minimum possible non-working travelled distance or the minimum non-productive time or the non-working cost), without overlaps or missed areas and avoiding all obstacles.
(e) Sub-fields sequence. It regards the determination of the sequence that the mobile unit visits the sub-fields given the access paths between them.

6.3 Results

An optimized route planning is presented under the criterion of optimization of operational costs. In operational costs, any cost is included that is directly or indirectly connected with the field operations application, e.g. fuel consumption, idle time costs, non-effective material cost etc.

In the following, the comparison between traditional and optimized route planning of field operations carried out by one unit is presented under the criterion of operational costs. In traditional route planning, the driver starts working in a block and moves to the next one only after the completion of the work in the first one. On the other hand, in optimized route planning, route planning can be mixed in the different blocks of the same field to minimize the operational costs. To examine the range of the size of agricultural machinery, four different machinery scenarios/cases are presented with the corresponding working width and turning radius (Table 6.1).

The corresponding four different field areas are:

- Field A: 6.01 ha
- Field B: 5.65 ha
- Field C: 5.70 ha
- Field D: 3.76 ha

The comparison of operational time in seconds for traditional and optimized route planning of field operations for the four fields examined is shown below in Fig. 6.1. Field efficiency is directly connected to the operational time; it can be defined as the ratio of the time a field machine is operating effectively to the total time that this machine is committed to the field operation (Bochtis et al. 2010b) given as a percentage.

Given that the average financial calculated cost per operational time (sec) is 0.0453 euros s^{-1}, the comparison of operational cost in euros for traditional and optimized route planning of field operations for the four fields examined can be extracted as shown below in Figs. 6.2, 6.3, 6.4 and 6.5. Among the four different field machinery scenarios, there is considerable divergence in operational cost from the smaller to the larger machinery and of course the cost is reduced by following the optimized route planning regardless.

Table 6.1 Characteristics of the machinery scenarios/cases

	Case A	Case B	Case C	Case D
Size	Small size	Small-medium size	Medium-large size	Large size
Working width (m)	1.5	3	3	6
Min turning radius (m)	3.5	3.5	5	5

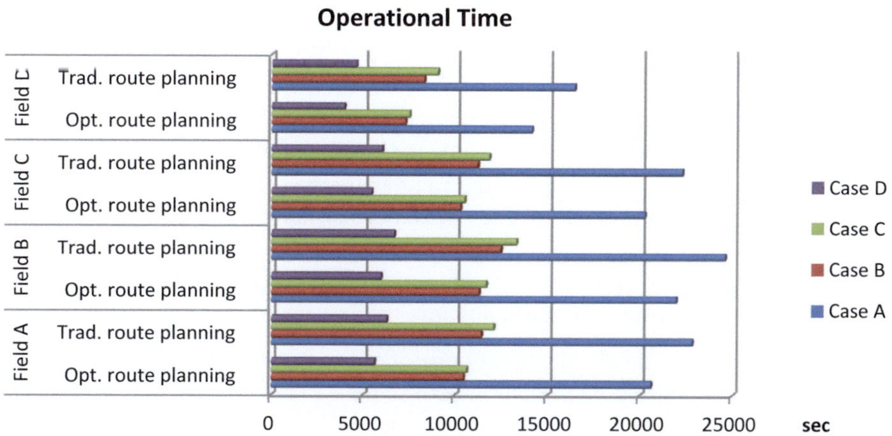

Fig. 6.1 The comparison between traditional and optimised route planning for operational time (s) in Fields A–D

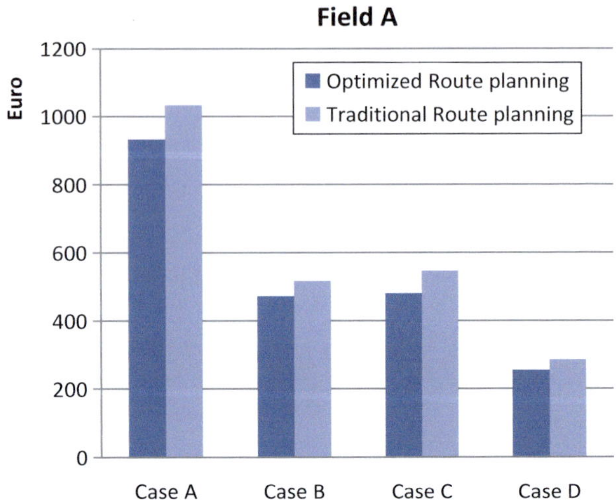

Fig. 6.2 The comparison between traditional and optimized route planning for operational cost (euros) in Field A

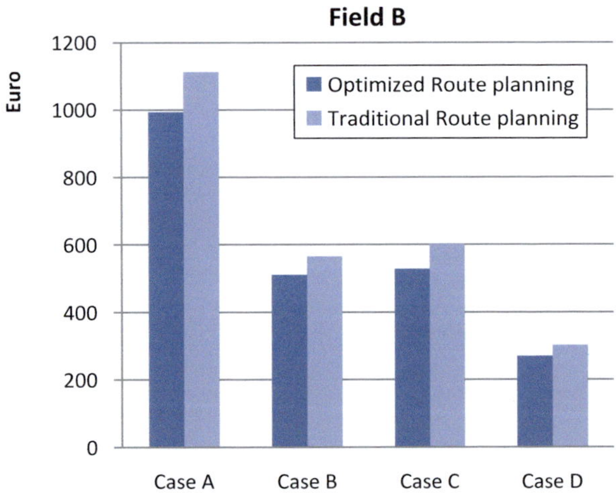

Fig. 6.3 The comparison between traditional and optimized route planning for operational cost (euros) in Field B

The comparison between the four fields presented above indicates the considerable savings achived by using optimized route planning. In Tables 6.2, 6.3, 6.4 and 6.5, these savings are presented for each field. Furthermore, these savings are shown graphically in Fig. 6.6.

Concerning the non-effective cost that comes from the non-effective time, the factors that have been taken into account are:

- Time of turning on headlands during operations at the main field
- Time of turning during operations at headlands
- Time of travelling from farm to field and back

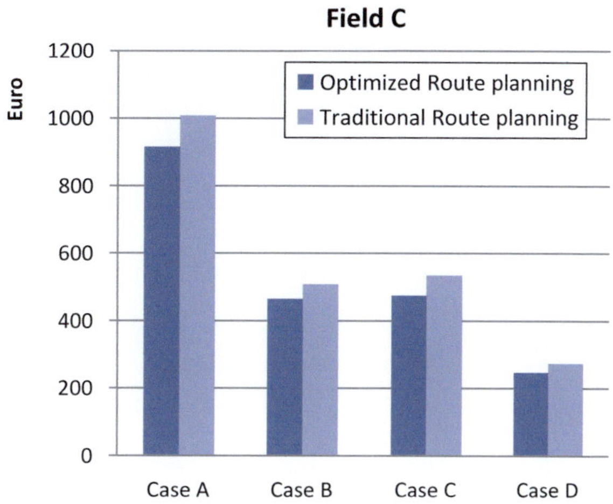

Fig. 6.4 The comparison between traditional and optimized route planning for operational cost (euros) in Field C

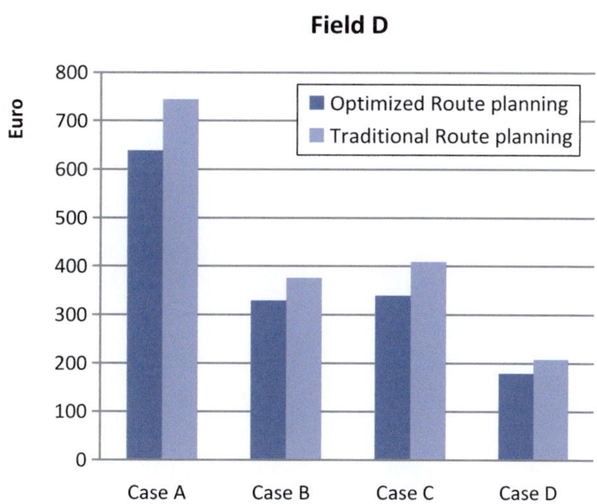

Fig. 6.5 The comparison between traditional and optimized route planning for operational cost (euros) in Field D

Table 6.2 Savings % in operational cost by using optimised route planning in Field A

	Working width (m)	Min turning radius(m)	Savings %
Case A	1.5	3.5	11
Case B	3		9
Case C		5	14
Case D	6		12

Table 6.3 Savings % in operational cost by using optimised route planning in Field B

	Working width (m)	Min turning radius(m)	Savings %
Case A	1.5	3.5	12
Case B	3		11
Case C		5	14
Case D	6		12

Table 6.4 Savings % in operational cost by using optimised route planning in Field C

	Working width (m)	Min turning radius(m)	Savings %
Case A	1.5	3.5	10
Case B	3		9
Case C		5	13
Case D	6		11

Table 6.5 Savings % in operational cost by using optimised route planning in Field A

	Working width (m)	Min turning radius(m)	Savings %
Case A	1.5	3.5	17
Case B	3		14
Case C		5	20
Case D	6		17

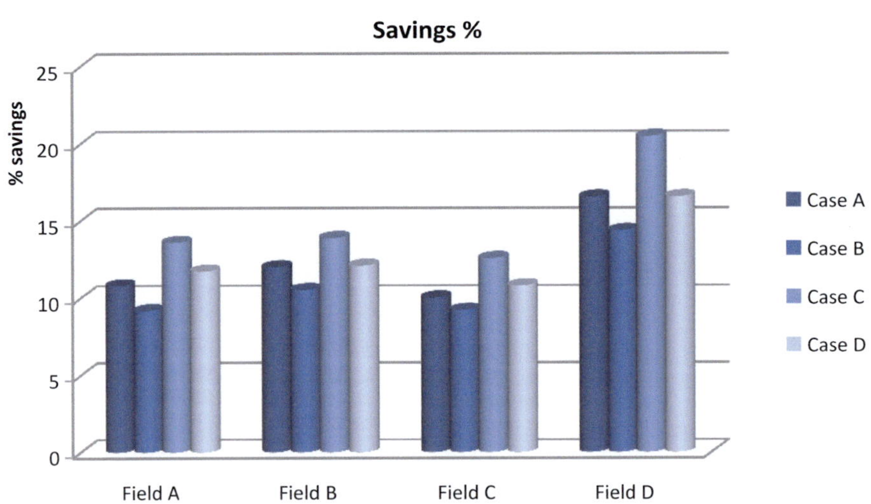

Fig. 6.6 Savings % in operational cost with optimised route planning for the four fields

Regarding the machine capacity the following factors should be taken into account:

- Machinery preparation time in the field before and after field operations without including daily services, lubrication and preparation for towing
- Machinery adjustment time
- Maintenance time (e.g. refueling)
- Operator's personal time

Concerning the implementation of the average of the above factors, a time delay of 1.65 min ha^{-1} (100 s ha^{-1}) was added to the operational time and an addition of 5% dedicated to personal breaks (Sørensen 2003; Sørensen and Nielsen 2005) (Tables 6.6, 6.7, 6.8, 6.9 and 6.10).

Table 6.6 Machine effective capacity for Field A

	Capacity (ha/h)		
	Traditional	Optimised	Increase %
Case A	0.95	1.05	11
Case B	1.90	2.07	9
Case C	1.79	2.03	14
Case D	3.44	3.84	12

Table 6.7 Machine effective capacity for Field B

	Capacity (ha/h)		
	Traditional	Optimised	Increase %
Case A	0.83	0.93	12
Case B	1.63	1.80	11
Case C	1.53	1.74	14
Case D	3.04	3.41	12

Table 6.8 Machine effective capacity for Field C

	Capacity (ha/h)		
	Traditional	Optimised	Increase %
Case A	0.92	1.01	10
Case B	1.83	2.00	9
Case C	1.74	1.96	13
Case D	3.40	3.77	11

Table 6.9 Machine effective capacity for Field D

	Capacity (ha/h)		
	Traditional	Optimised	Increase %
Case A	0.82	0.96	17
Case B	1.63	1.87	14
Case C	1.50	1.81	20
Case D	2.95	3.44	17

Table 6.10 The four machinery cases regarding the power engine

	Size	Power (kw)
Case A	Small	30
Case B	Small-medium	60
Case C	Medium-large	90
Case D	Large	120

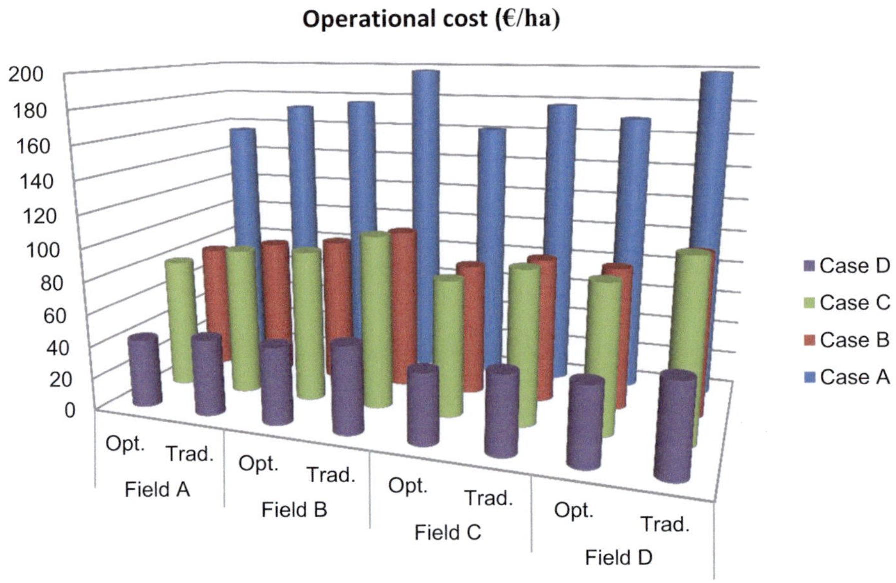

Fig. 6.7 Operational cost in euros/ha for both optimised (opt.) and traditional (trad.) route planning for the 4 fields

In Fig. 6.7 below, the operational cost in euros per ha is presented for both traditional and optimised route planning in the 4 fields examined.

To calculation fuel consumption, the equation given at Agricultural Machinery Management Data, D497.4 (ASAE 2003) was used:

$$2.64X + 3.91 - 0.203\sqrt{738X + 173} \text{ in } \frac{l}{kw \times h},$$

where X is the ratio of equivalent Power Take-Off (PTO) power required by an operation to the maximum available from the PTO. To evaluate an 'average' field operation, X was set to 75% while operating. During turnings X was set to 0% (the PTO was off).

The results that correspond to the above mentioned method of optimisation in fuel consumption is shown in Tables 6.11, 6.12, 6.13 and 6.14 for the examined field areas including the corresponding fuel savings in each case.

In Fig. 6.8, the fuel consumption in litres ha^{-1} is shown for machinery scenarios A-D for the four fields. The introduction of optimised route planning compared to the traditional ones results in reduced fuel consumption for different field sizes and variable field machinery equipment (Cases A–D). Furthermore, in Fig. 6.9, the cor-

Table 6.11 Fuel consumption in liters for traditional and optimised route planning for Field A

	Fuel consumption (l)		
	Traditional	Optimized	Fuel savings %
Case A	90.77	81.91	11
Case B	90.89	83.25	9
Case C	144.29	127.01	14
Case D	100.18	89.66	12

Table 6.12 Fuel consumption in liters for traditional and optimised route planning for Field B

	Fuel consumption (l)		
	Traditional	Optimized	Fuel savings %
Case A	97.79	87.30	12
Case B	99.32	89.85	11
Case C	158.79	139.42	14
Case D	106.49	94.99	12

Table 6.13 Fuel consumption in liters for traditional and optimised route planning for Field C

	Fuel consumption (l)		
	Traditional	Optimized	Fuel savings %
Case A	88.59	80.50	10
Case B	89.32	81.78	9
Case C	141.14	125.36	13
Case D	95.98	86.62	11

Table 6.14 Fuel consumption in liters for traditional and optimised route planning for Field D

	Fuel consumption (l)		
	Traditional	Optimized	Fuel savings %
Case A	65.44	56.14	17
Case B	66.08	57.78	14
Case C	107.88	89.53	20
Case D	73.11	62.73	17

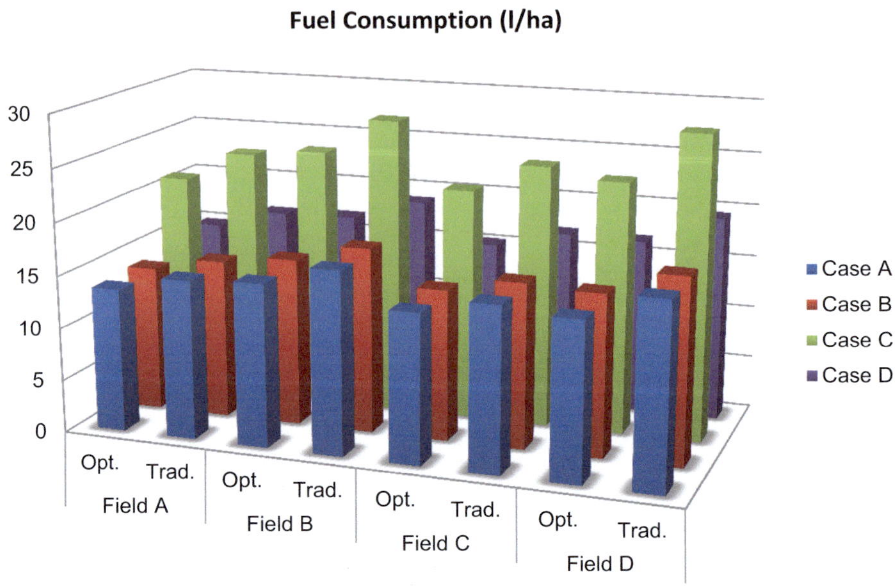

Fig. 6.8 Fuel consumption in litres/ha for both optimised (opt.) and traditional (trad.) route planning for the 4 fields

responding cost in euros from fuel consumption is presented given that the mean diesel fuel price throughout Europe at present is around 1.05 euro litre^{-1}.

6.4 Discussion

From the results above, we can indicate that the operational cost savings by using the optimised route planning are: for Field A the cost savings vary from 9 to14% with the best one in case C, in Field B the cost savings vary from 11 to 14% with the best one in case B, in Field C the cost savings vary from 9 to 13% with the best one in case C and finally in Field D the cost savings vary from 14 to 20% with the best one in case C. Furthermore, an optimised route planning for the Fields A, B, C and D will result in an increase in field machinery capacity of 9–14%, 11–14%, 9–13% and 14–20%, respectively. The results are similar for the reduction in fuel consumption and in fuel cost per ha for the four Fields. In conclusion, there is an immediate connection between the increase in field machinery capacity, the savings in operational cost and the savings in fuel cost.

The results show that in examples with very small or large machinery (case A and case D with corresponding working widths of 1.5 and 6 m, respectively, and corresponding minimum turning radii 3.5 and 5 m, respectively) route planning optimisation provides significant operational cost savings compared to non-optimised routings, thus case A and D provide the largest savings in terms of costs.

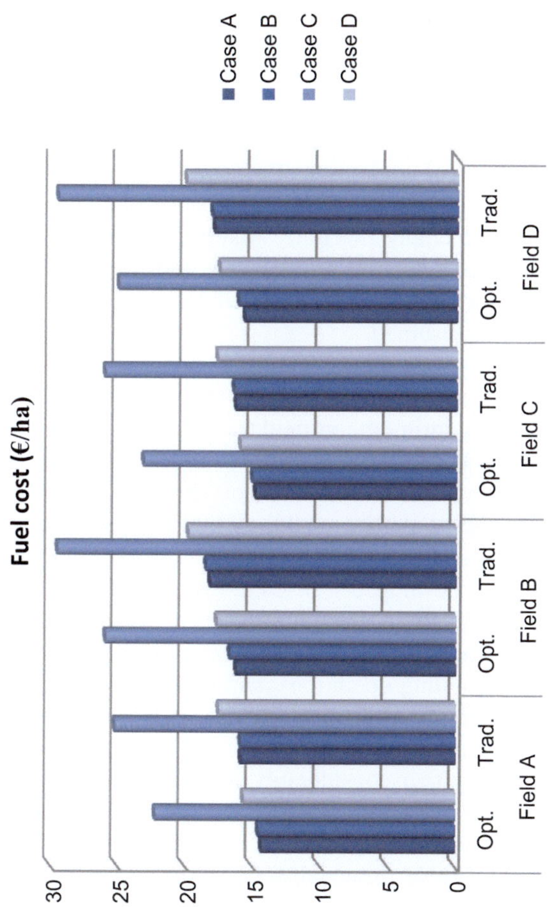

Fig. 6.9 Fuel cost in euros/ha for both optimised (opt.) and traditional (trad.) route planning for the 4 fields

This issue should be studied more extensively, given that even though large machinery is directly connected with large working widths and corresponding large field capacities, nevertheless, the largest increase in field capacity is not presented in case D. The same issue, in reverse, is seen with very small machinery that connects immediately with a small turning radius.

Beyond the benefits of CTF, as mentioned in the introduction section, also CTF has some drawbacks mainly derived from the constraints imposed in the paths that an agricultural machine can follow. This is more evident in the case of material handling operations such as organic fertilizing (i.e. manure application) where there is the need for in-field transport of the machine to refill. When there is no coordination between the length of the fieldwork track and the driving distance corresponding to the application of a full load, the traffic restrictions of the CTF system do not allow for in-field turning of the machine and the machine must drive empty along the remaining part of the field work track. This non-working travelled distance can increase further when the entry and exit locations of the field are not located in the travelling direction of the empty machine. Analogous situations occur when the machine comes back to the field after refilling (e.g. fertilizers) and must travel over a part of a field work track without applying fertilizers in order to reach the location where application may be resumed.

To that effect, in the case of material handling operations, the implementation of the CTF system can reduce the field efficiency of these operations because of the non-productive time spent during the in-field transport. Consequently, planning for field coverage in material handling operations under the CTF path restrictions becomes critical in order to reduce the trade-off in field efficiency. However, the interrelations between the properties of the motion sequences of the agricultural machinery and the configurations of the CTF layout are extremely complex.

Bochtis et al. (2009a) developed a discrete-event model for the prediction of travelled distances of agricultural machines operating in material handling operations in a CTF system. It was proved that the key factor that affects field efficiency in the case of material handling operations, and specifically in the case of organic fertilizing, is the in-field transport distance. Based on experimental results (Bochtis et al. 2010b) in two fields, it was shown that the implementation of CTF instead of the conventional traffic system considerably increases the in-field transport distances. Specifically for the operations examined in two fields, the estimated increase in the transport distance was 47.82% and 24.54% resulting in a reduction of field efficiency of 7.41% and 4.68%, respectively. In another study based on a simulation model, it was also shown that the implementation of the CTF system increases the operational time by up to 5%, resulting in a decrease in the field efficiency in the range of 11.52–8.25% (Hameed et al. 2012).

The route optimization described improves the field efficiency of material handling operations by minimizing the various non-productive travelled distances. However, the prerequisite for this minimization is the operational analysis for identifying the activities, the actions and their interconnections that contribute to the reduction of the efficiency (Jensen et al. 2015b). An approach on this minimization problem has been presented recently by Jensen et al. (2015a) based on the state-space search technique where the solution of the problem is a sequence of predefined driving actions that transform the initial state to a goal state under the criterion of minimizing the in-field non-working travelled distance. In the specific approach, the sequence of the working tracks is optimised in a post-process where the travelling salesman problem methodology (Hoffman et al. 2013) is applied to minimize the non-working distance travelled while turning at headlands. Results showed that by implementing the travelling salesman approach for the field coverage optimisation the savings achieved in the non-working travelled distance (including both in-field transports and headland turning) amounted to 15.7%, 43.5% and 23% for the three fertilizing operations examined. These numbers correspond to savings in the total travelled distance of 5.8%, 11.8% and 11.2%, respectively.

Another critical factor that can affect the field efficiency of the operations in the CTF system is the direction of the field work tracks in relation to the field shape, since a long-term configuration must be determined. When considering the field work tracks direction, however, the entire set of operations executed in the field has to be taken into consideration. Bochtis et al. (2010a) presented an approach to estimate the operational machinery costs in the CTF system based on a number of sub-models to evaluate the consequences, in terms of machinery performance, for different potential driving directions in a field when establishing the permanent fieldwork tracks. The approach takes into account the non-working distance travelled during the headland turnings, the in-field travelling distance for the case of the material handling operations, and moreover, the cost of lost material resulting from overlaps in the area covered. The most important result of this study was the conclusion that the rule prevailing in the conventional traffic system that the optimal driving direction is the one parallel to the longest edge of the field does not apply in the CTF system. For example, in the case of a specific field the annual cost of machinery operation decreased by 9% when the direction of the field work tracks was the one parallel to the shortest edge of the field. This is a result of various factors including the area overlapped in spraying and seeding, the unloading times in harvesting, the in-field transport in material handling operations and the non-working distance during headland turnings. Overall, the benefits from optimized route planning provide cost savings in the range of 9–20% and increased field machinery capacity also ranging from 9 to 20%. In the case of material handling, implementation of the CTF system increases operational time by up to 5% and reduces the field efficiency from 8 to 11%. The benefits mentioned can be obtained by implementing the route planning software, either as a manual decision support system or directly coupled to the auto guidance system.

6.5 Conclusion

Navigation systems and field machinery automation systems such as auto steering guidance and field management systems such as Controlled Traffic Farming that are used in modern agriculture have been assessed in this chapter for different machinery configurations. Traditional and optimised route planning systems have been compared in four different field areas. Considerable savings in operational and in fuel costs of up to 20% in optimised route coverage system were observed. Specifically, the cost savings ranged from 9 to 20% with an associated increase of the machinery capacity in the same range. Disadvantages from introducing CTF in the case of material handling include up to 5% increase in operational time and a reduced field efficiency of 8–11%. The benefits mentioned can be obtained by implementing the route planning software, either as a manual decision support system or directly coupled to the auto guidance system. These results could be examined further through extended research and experimentation in route planning design not only under the criterion of reduction of operational cost, but also under other environmental criteria, such as reduction of energy consumption and or a reduction of CO_2 emissions. The benefits that are quite significant, as described above, and thus this solution could play and important role in the criterion of minimizing the operational cost and time.

References

ASAE S (2003) ASAE D 497.4 FEB03 Agricultural machinery management data American Society of Agricultural Engineers

Batte MT, Ehsani MR (2006) The economics of precision guidance with auto-boom control for farmer-owned agricultural sprayers. Comput Electron Agric 53:28–44

Bochtis D (2008) Planning and control of a fleet of agricultural machines for optimal management of field operations. Aristotle University, Greece

Bochtis DD, Sørensen CG (2009) The vehicle routing problem in field logistics part I. Biosyst Eng 104:447–457

Bochtis DD, Sørensen CG (2010) The vehicle routing problem in field logistics: part II. Biosyst Eng 105:180–188

Bochtis DD, Vougioukas SG (2008) Minimising the non-working distance travelled by machines operating in a headland field pattern. Biosyst Eng 101:1–12

Bochtis DD, Sørensen CG, Jørgensen RN, Green O (2009a) Modelling of material handling operations using controlled traffic. Biosyst Eng 103:397–408

Bochtis DD, Vougioukas SG, Griepentrog HW (2009b) A mission planner for an autonomous tractor. Trans ASABE 52(52):1429–1440. 10.13031/2013.29123

Bochtis DD, Sørensen CG, Busato P et al (2010a) Tramline establishment in controlled traffic farming based on operational machinery cost. Biosyst Eng 107:221–231

Bochtis DD, Sørensen CG, Green O et al (2010b) Effect of controlled traffic on field efficiency. Biosyst Eng 106:14–25

Bochtis DD, Sørensen CG, Busato P, Berruto R (2013) Benefits from optimal route planning based on B-patterns. Biosyst Eng 115:389–395

Bochtis DD, Sørensen CGC, Busato P (2014) Advances in agricultural machinery management: a review. Biosyst Eng 126:69–81. https://doi.org/10.1016/j.biosystemseng.2014.07.012

Bochtis D, Griepentrog HW, Vougioukas S et al (2015) Route planning for orchard operations. Comput Electron Agric 113:51–60

Chesworth W (ed) (2008) Encyclopedia of soil science. Springer, Dordrecht

Engfeldt A (2005) Network RTK in northern and central Europe. LMV-Rapport 2005:5 - ISSN 280-5731 Lantmäteriet, Gävle, Sweden

Gan-Mor S, Clark RL (2001) DGPS-based automatic guidance – implementation and economical analysis. In: 2001 Sacramento, CA July 29–August 1,2001. American Society of Agricultural and Biological Engineers, St. Joseph, p 1

Hameed IA, Bochtis DD, Sørensen CG, Vougioukas S (2012) An object-oriented model for simulating agricultural in-field machinery activities. Comput Electron Agric 81:24–32. https://doi.org/10.1016/j.compag.2011.11.003

Hoffman KL, Padberg M, Rinaldi G (2013) Traveling salesman problem Encyclopedia of operations research and management science. Springer, New York, pp 1573–1578

Jensen MF, Bochtis D, Sørensen CG (2015a) Coverage planning for capacitated field operations, part II: optimisation. Biosyst Eng 139:149–164. https://doi.org/10.1016/j.biosystemseng.2015.07.002

Jensen MF, Nørremark M, Busato P et al (2015b) Coverage planning for capacitated field operations, part I: task decomposition. Biosyst Eng 139:136–148

Kroulík M, Kvíz Z, Kumhála F et al (2011) Procedures of soil farming allowing reduction of compaction. Precis Agric 12:317–333. https://doi.org/10.1007/s11119-010-9206-1

Lawson LG, Pedersen SM, Sørensen CG et al (2011) A four nation survey of farm information management and advanced farming systems: a descriptive analysis of survey responses. Comput Electron Agric 77:7–20. https://doi.org/10.1016/j.compag.2011.03.002

Sørensen CGG (2003) Workability and machinery sizing for combine harvesting. CIGR J AE Sci Res Dev V:Manuscript PM 03:003

Sørensen CG, Bochtis DD (2010) Conceptual model of fleet management in agriculture. Biosyst Eng 105:41–50

Sørensen CG, Nielsen V (2005) Operational analyses and model comparison of machinery systems for reduced tillage. Biosyst Eng 92:143–155. https://doi.org/10.1016/j.biosystemseng.2005.06.014

Sørensen CG, Fountas S, Nash E et al (2010) Conceptual model of a future farm management information system. Comput Electron Agric 72:37–47. https://doi.org/10.1016/j.compag.2010.02.003

Tillett ND (1991) Automatic guidance sensors for agricultural field machines:a review. J Agric Eng Res 50:167–187. https://doi.org/10.1016/S0021-8634(05)80012-5

van Zuydam RP, Sonneveld C (1994) Test of an automatic precision guidance system for cultivation implements. J Agric Eng Res 59:239–243. https://doi.org/10.1006/jaer.1994.1082

Chapter 7
Profitability of Controlled Traffic in Grass Silage Production

Hans Alvemar, Hans Andersson, and Hans Henrik Pedersen

Abstract Controlled traffic farming (CTF) systems aim to reduce soil compaction by restricting machinery field traffic to permanent traffic lanes. Soil compaction and field traffic from heavy machinery is known to affect crop growth negatively. Grass-clover silage production is generally associated with intensive field traffic, resulting in reduced silage clover content. If CTF can increase yield and clover content in grass-clover leys, this would reduce the need for grain and expensive protein concentrate in dairy cow feed rations. However, the CTF system often involves changes to machinery systems. This cost must be examined when evaluating the profitability of converting to CTF. A mixed integer programming model has been developed to evaluate the potential profitability of CTF in a dairy farm context. Existing field trial data were used to calculate the expected yield outcome of CTF, based on reductions in trafficked area. The results revealed that CTF increased profitability by up to €50 per hectare when silage yield or quality increased with CTF. Total machinery costs are likely to increase when converting to CTF, but variable machinery costs are likely to decrease. Overall, if CTF increases yield or silage quality, the system is profitable despite the major investment required. This chapter addresses agronomic as well as practical aspects of CTF systems for grass silage production.

Keywords Controlled traffic farming • Auto steering • Grass silage • Profitability • Integer programming

H. Alvemar (✉)
SOYL Sverige AB, Skara, Sweden
e-mail: hans.alvemar@soyl.se

H. Andersson
Swedish University of Agricultural Sciences, Uppsala, Sweden
e-mail: Hans.Andersson@slu.se

H.H. Pedersen
Aarhus University, Aarhus, Denmark
e-mail: hansh.pedersen@eng.au.dk

S.M. Pedersen, K.M. Lind (eds.), *Precision Agriculture: Technology and Economic Perspectives*, Progress in Precision Agriculture,
https://doi.org/10.1007/978-3-319-68715-5_7

7.1 Introduction

Grass leys are an important part of preserving soil structure and managing pests in a crop rotation. Intensive livestock production units with high yielding dairy cows require large volumes of high quality grass silage. This chapter examines the possibilities to increase yields and improve quality of grass silage using controlled traffic farm management systems. Soil compaction and mechanical plant damage caused by field traffic affects both dry matter yield and the botanical composition of mixed grass clover leys (Hansen 1996). The aim of using CTF is to reduce the effects of random soil compaction and mechanical plant damage by confining all field traffic to the least possible area of permanent traffic lanes. This study focuses on evaluating the profitability of CTF in grass silage production.

7.1.1 Soil Compaction and Field Traffic

According a to a large number of previous studies (Chamen 2011) soil compaction is, in many cases, known to have a negative effect on several aspects of crop production. Even though the research topic on crop yields and soil compaction dates back to the pre-mechanized agricultural production, this problem becomes more relevant as agricultural machinery becomes larger. Soil compaction is a major concern in modern agriculture because of increased farm size and larger machinery, which may lead to persistent subsoil compaction (Håkansson and Reeder 1994; Keller and Arvidsson 2004). Large-scale grass silage production includes numerous machinery passes during the growing season, including several passes with total machinery weights exceeding 30 tonnes. Hence, the possibility to minimize and systemize machinery traffic in permanent traffic lanes provides an opportunity to reduce soil compaction and crop damage in grass silage production.

Soil structure as well as grass and clover plants are sensitive to machinery traffic during wet conditions. Low porosity compacted soils are characterized by low water infiltration capacity (Raper 2005), which increases the top soil water content, which subsequently increases the risk of soil compaction during field traffic. Chyba et al. (2014) state that non-compacted soil has 4–5 times greater water infiltration rates than soil compacted by agricultural machinery. Even compaction caused by cattle hooves has proved to decrease water infiltration rates by 80% (Chyba et al. 2014).

Tillage has for many years been the countermeasure for achieving desirable soil structure on compacted soil which have suffered from heavy machinery traffic. Arndt and Rose (1966) stated that excessive field traffic needs increased tillage. Hence, the use of a controlled traffic system where all field traffic is confined to permanent traffic lanes should reduce the need for tillage. Furthermore, tillage operations can be confined to the wheel tracks, thus reducing the energy input required for tillage. Cereal dominated cropping systems are characterized by intensive tillage to repair soil structure and damage from heavy machinery traffic. However, in a perennial grass ley the possibilities to repair soil structure are limited. In cropping systems where a grass ley is established by under-sowing in spring cereals there is a high risk of causing substantial damage to the soil and to the grass crop during the cereal harvest operations, with no possible measures to repair damage from previous machinery passes. Hence, the yield potential of the grass ley may be reduced before the first cut. In addition to this, the grass silage harvest system includes multiple machinery passes from several cuts during the years of the ley. Consequently grass silage production induces heavy machinery traffic without the possibility to repair soil structure during the years of ley.

7.1.2 Field Trials – Yield and Botanical Composition

A summary of results from field trials where herbage yields from grass leys exposed to machinery traffic are compared to non-trafficked plots indicates major yield reductions from machinery traffic (Table 7.1). Hansen (1996) compared trafficked and non-trafficked plots during 3 years of ley in three cuts per year harvest system. The results display an average of 27% in yield reduction from machinery traffic. Studies conducted over several years provide more relevant results compared to one-cut trials when examining the profitability for dairy production farms. The results presented by Douglas et al. (1992) reveal substantial differences in herbage yield from zero traffic systems in years two, three and four.

Table 7.1 Summary of literature studies, effects of machinery traffic on grass/clover yields

Study	Yield decrease (%)	Grass crop	Soil type	Location
Douglas and Crawford (1991)	32	Ryegrass	Clay loam	Scotland
Douglas et al. (1992)	13	Ryegrass	Clay loam	Scotland
Elonen (1986)	8–68	N/A	Clay loam	Finland
Frame (1982)	11–36	Red clover	N/A	Scotland
Frost (1988)	9–13	Ryegrass	Clay loam – Sandy clay loam	Northern Ireland
Hansen (1996)	27	Grass/clover	Sandy loam	Norway
Håkansson et al. (1990)	9	Grass/clover	Various	Sweden
Jorajuria et al. (1997)	74	Grass/clover	Silty loam	Argentina
Jørgensen et al. (2009)	4.6–23	Grass/clover	N/A	Denmark
Rasmussen and Møller (1981)	21–54	Ryegrass & Grass/clover	Sandy loam & silty loam	Denmark

The results from soil compaction trials focusing on cereal crops on clay soils support the results presented above displaying yield reductions ranging from 15 to 26% (Chamen et al. 1992; Chamen and Longstaff 1995; Dickson and Ritchie 1996).

7.1.3 Feed Rations and Silage Quality Variations

One of the potential benefits of controlled traffic farming in grass clover leys is to increase forage quality and protein contents. Hansen (1996) found that soil compaction could alter both herbage yields and the botanical composition of grass clover leys. Hansen's results display a reduced red clover content by 8% after 3 years of compaction compared to non-compacted soils, which supports Frame (1982) who stated that red clover is particularly sensitive to soil compaction. Clover for ruminants is nutritionally superior to grasses with respect to protein, mineral content and feed intake capacity (McDonald et al. 2011). Hence, the botanical composition of grass clover leys is important in dairy production. To evaluate the potential profitability of increased forage quality with CTF a feed ration perspective must be included when evaluating the total benefits of grassland controlled traffic farming.

The choice of grass harvest equipment may also have an effect on silage quality. However, in this study the choice of machinery is made solely considering the trafficked field area in a CTF system. The machinery systems evaluated in this study include different silage chopping methods. Previous studies comparing rotor cutter wagons and precision chopping wagons showed no differences in silage fermentation, yet the rotor cutter wagon increases loading and ensiling capacity (Lingvall and Arvidsson 2005; Lingvall and Knicky 2008). Therefore, the choice of chopping method is not considered to have an effect on silage quality large enough to affect the choice of machinery when converting to CTF. However, Nadeau et al. (2014) argue, that to maintain silage quality, a mower without crimper should be not used when not using a rotor cutter wagon, but may be used if using a precision chopper or self-propelled forage harvester. Hence, this study applies the combinations of mower with crimper and rotor cutter wagon, or rotary mower and precision chopper.

7.2 Machinery System Design

The design of an optimal machinery system is unique for each farm. Previous studies present optimization models for machinery size selection (Søgaard and Sørensen 2004). However, results for the optimal machinery size often indicate that various machinery widths should be used to optimize the available tractor power. A controlled traffic machinery system is designed to minimize the field area covered by machinery traffic, referred to as trafficked area, by organizing all machinery traffic into permanent traffic lanes. By using a standardized machinery working width, known as a module width, or multiples of the module width, all field traffic can be

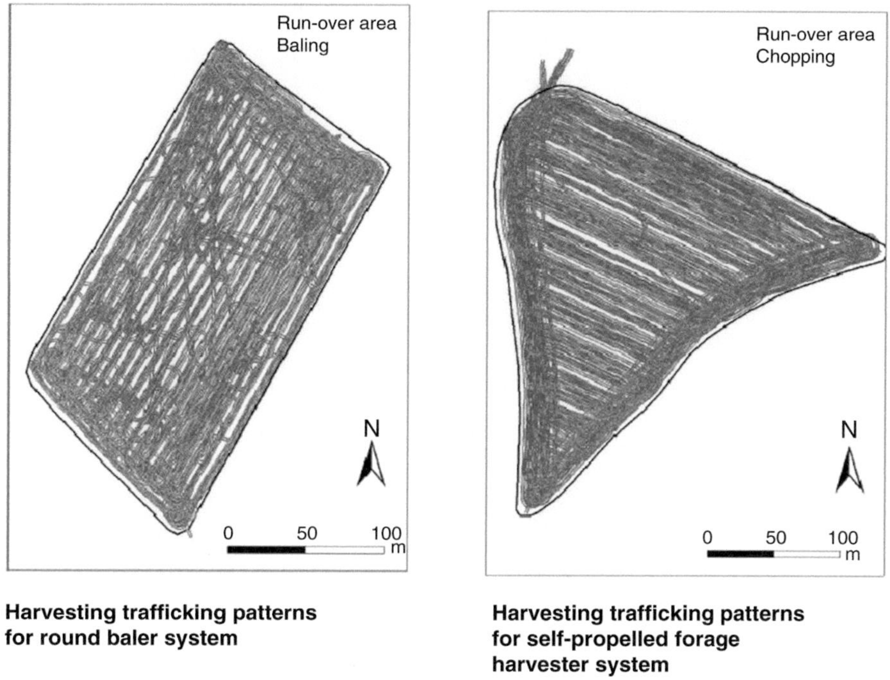

Fig. 7.1 Field traffic map from one silage cut with mower, rake, tedder and self-propelled forage harvester, trafficked areas over 60% (Kroulik et al. 2014)

confined into permanent traffic lanes (Tullberg et al. 2003). Hence, reducing trafficked area and soil compaction may at a first glance seem to require suboptimal machinery use. However, soil structure and fertility are of uttermost importance for long-term profitability in agricultural production.

Agricultural machinery operated in a conventional random traffic farming system may cause poor soil structure, covering large areas with machinery traffic. Kroulik et al. (2014) measured the traffic intensity from two common silage harvest systems and found trafficked field areas of 64% from one silage cut. The traffic pattern displays random distribution of field traffic causing great variation in soil compaction and mechanical plant damage and thus in the silage quality in forthcoming cuts (Fig. 7.1).

7.2.1 *Traffic Management Systems*

The CTF system includes three basic principles: first, the machinery is set up in working width modules, second, the wheel track gauge needs to be matched to minimize the tracked area and third, it is crucial to use auto steering and the global

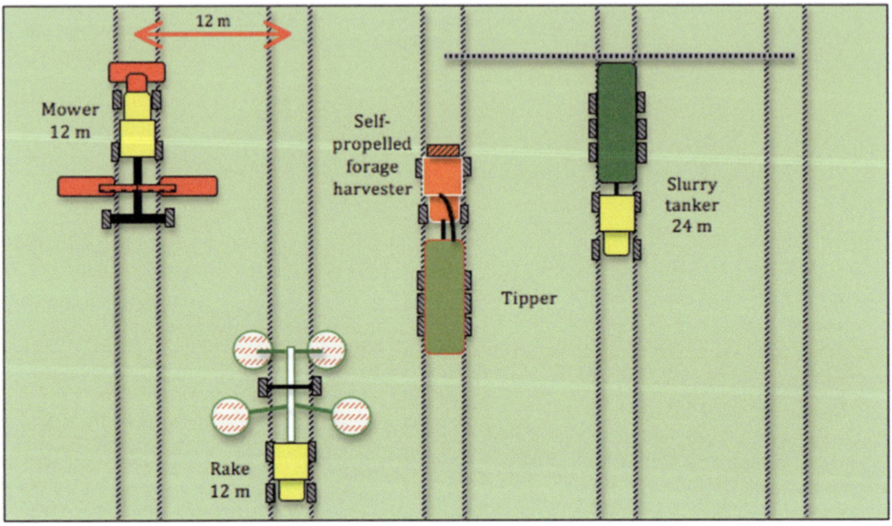

Fig. 7.2 CTF system for grass silage

navigation satellite system (GNSS) with real time kinematics (RTK) correction signal to ensure that all traffic is confined to the traffic lanes (Webb et al. 2004).

These three principles make it possible to set up and maintain a controlled traffic system as illustrated in Fig. 7.2. Using a machinery traffic management system reduces variation in soil compaction and mechanical plant damage. The controlled traffic farming system confines all machinery traffic to permanent traffic lanes to reduce the tracked field area.

7.2.1.1 Controlled Traffic in Grass Ley

The machinery operations associated with grass silage harvest and slurry or manure application often induce high wheel loads with intensive traffic patterns, as presented by Kroulik et al. (2014). Controlled traffic systems can be used only in grassland operations, whereas machinery operations in arable crops follow a random traffic system. The use of tillage and seeding tools at 8 m working width or larger require powerful and heavy tractors. However, for grass harvest, the relatively low power requirements for mowers, rakes and rotor cutter or chopping wagons allow for relatively large working widths, such as eight to 12 m, with standard tractors, often less than 250 HP power. Hence, a change to controlled traffic farming solely in the grass harvest may be possible without investing in large tractors.

The common CTF-system in grass requires all tractor passes to be driven astride the grass swaths at the centre of the full working width. Consequently, the system necessitates a butterfly mower and a centre-swath twin- or four-rotors rake. Furthermore, silage chopping and collection is best practiced with a chopper wagon or a self-propelled forage harvester towing a silage trailer where the majority of all traffic is confined to the permanent traffic lanes. Slurry, manure and mineral fertilizer applications should be included in the system design and selection of a common CTF working width.

When choosing a base module width, the principle of least possible tracked area from least possible investment is a sound approach. In theory, any module width can be applied to a CTF system. However, practically there are some working widths that are easy to use when restricted by the standard machinery available in the market. Naturally, the size and shape of field as well as the total farmed area will also have an effect on which working width is suitable. As mentioned previously, the grass CTF system generally includes harvest of silage, mineral fertilizer, manure and slurry applications (Fig. 7.3).

Practically controlled traffic systems for grass silage can be designed using 8, 9 and 12 m, and occasionally with 10 and 11 m module widths. In this example three systems, using 8, 9 and 12 m module widths are presented. The three systems in Table 7.2 will all result in approximately 20% trafficked area, but with different grass preparation and collection methods.

The three systems are based on a butterfly mower with or without crimper, a twin- or four-rotor centre swath rake and a silage chopper wagon or a self-propelled

Fig. 7.3 A CTF system for grass silage (12-m width) at Moosegården in Denmark

Table 7.2 Machinery systems used for the profitability evaluation

	Mower	Rake	Silage handling
RTF	Front + rear mower, 6 m working width	Two rotor rake, side swath placement	Rotor cutter wagon
CTF 8	Tractor mounted butterfly mower with crimper 8.3 m working width	Two rotor rake, centre swath	Rotor cutter wagon
CTF 9	Tractor mounted butterfly mower with crimper 9.5 m working width	Two rotor rake, centre swath	Rotor cutter wagon
CTF 12	Towed butterfly mower without crimper, 12.3 m working width	Four rotor rake, centre swath	Self-propelled forage harvester

Controlled Traffic Farming (CTF) and Random Traffic Farming (RTF)

Fig. 7.4 Slurry tanker (14.5 m width) using a trailing shoe in CTF

forage harvester with associated tractors and trailers. Furthermore, the systems include a fertilizer spreader, roller and a slurry tanker.

The slurry tanker may use a spreader plate, a boom or incorporating equipment such as a trailing shoe (Fig. 7.4). A suitable spreading width for the slurry tanker is a double or triple base module width, i.e. 16 or 24 m in an 8-m CTF system. For easy machinery set-up in the field, a triple module width is preferred since this allows for full workability all the way to the field edge without any traffic lane adjustments. Where a double module width is used, either the first pass with the slurry tanker or the mower must be adjusted with one half of the base module width to achieve the least possible trafficked area where all traffic will be confined to the traffic lanes. If the slurry is spread with an incorporating tool it is common to use the same working width as the base module.

Controlled traffic farming is best practiced with tractor auto steering equipment with RTK precision. The possibility to reduce the investment when converting to CTF by minimizing the total number of auto steering equipped tractors is crucial for small and medium size farm units. When using the CTF system in grass silage production all tractors do not necessarily need auto steering equipment. If the tractor used for mowing is using an RTK based auto steering system, this will ensure accuracy and repeatability of grass swaths. Both the rake and silage chopping passes then follow the swath, hence these tractors do not necessarily require auto steering or may use a less advanced system than RTK. However, the tractor used for spreading slurry or manure should use RTK accuracy since this is the most damaging operation in terms of soil compaction.

7.3 Problem Definition

As stated in the introduction, this study analyses the use of heavy machinery traffic in grass silage production and the reduction in yields from compacted soil and mechanical plant damage. Controlled traffic farming systems propose a solution to minimize yield reductions from machinery traffic by confining all machinery traffic to the least possible area of permanent traffic lanes. However, the use of a controlled traffic machinery system may require investments to adapt the machinery system into CTF.

Scandinavia and Northern Europe often experience heavy rainfall (EEA 2015) during the grass harvest season. These weather conditions enhance the problems of soil compaction with intensive machinery traffic in grass silage production. Therefore, the aim of this study was to evaluate the profitability of controlled traffic farming systems in grass silage production in a dairy farm context. The study compares three CTF machinery systems to reduce the trafficked field area compared with a random traffic farming (RTF) scenario. Furthermore, three alternative scenarios of CTF in terms of silage quality are evaluated and addressed by comparing cost of feed rations.

The farm profits are modelled on a 300 ha dairy operation, housing a herd of 300 milking cows. The normal herbage yield for grass silage is assumed to be 8000 kg DM ha^{-1}. This is a very specialized diary production unit and is not to be considered an average farm. However, statistics show an increasing amount of large scale specialized dairy production units. The average herd size in Scotland was 224 cows in 2015 (AHDB 2016) and in Sweden 25% of all dairy cows were in herds over 200 cows (SJV 2016). The CTF system is interesting from a dairy perspective because it might offer not only increased herbage yield but also increased silage quality from clover protein, which may partly replace expensive compound feeds.

7.4 A Model for Economic Evaluation of CTF

This study was done with a programming model based on linear relations where mixed integer programming is used. The model aims to maximize the total farm profit depending on the distribution of crops, machinery system and feed rations for livestock.

The profitability of a CTF conversion for grass silage production is subject to the following factors: Machinery investment required to form a CTF machinery system and yield outcome, which in turn depends on the reduction of the area subject to machinery traffic.

The machinery investment scenarios in this study with CTF conversion follow the principle of least possible trafficked area with least possible investment cost. This implies that the profitability of CTF depends on finding a machinery system where a major reduction in trafficked field area can be achieved with minimal investments. However, this principle must not induce any negative changes to silage quality from reduced machinery capacity or poor silage handling procedures. If the CTF system is predicted to affect the silage quality this should be evaluated in economic terms.

The yield potential of CTF in grass silage production is the main incentive for a CTF conversion. In cereal production the incentives for converting to a controlled traffic system may be more diverse in terms of reduced fuel consumption and better crop establishment in direct drill systems. However, in grass silage production increases in terms of yield and quality is the major motivation for CTF.

7.4.1 Herbage Yield and Silage Quality

The yield potential of a CTF system is calculated based on the machinery systems trafficked area and the expected yield reduction from soil compaction or mechanical plant damage. Using the method from Pedersen and Novak (2013), the trafficked area for each system is calculated based on actual machinery measurements.

To calculate the theoretical no-traffic yield, results from Hansen (1996) are used to determine the relative yield reduction on trafficked and non-trafficked soils. With the potential no-traffic yield from Eq. (7.1) the yield for the CTF system is calculated based on the trafficked area of the CTF system in Eq. (7.2).

$$X_Y = \frac{Y_{\mathrm{RTF}}}{A_{\mathrm{T}} \cdot \left(1 - Y_{\mathrm{D}}\right) + A_{\mathrm{NT}}} \tag{7.1}$$

$$Y_{\mathrm{CTF}} = X_Y \cdot \left(A_{\mathrm{T}} \cdot \left(1 - Y_{\mathrm{D}}\right) + A_{\mathrm{NT}}\right) \tag{7.2}$$

where Y_{RTF} is yield with random traffic, Y_{CTF} is yield with controlled traffic, $_{\mathrm{Y}}$ is theoretical no-traffic yield, Y_{D} is yield depression from field traffic, A_{T} is trafficked area and A_{NT} is non-trafficked area.

The theoretical no-traffic yield X_{Y} is based on knowledge of the relations between trafficked and non trafficked plots, where the farm machinery system trafficked area is used to determine the theoretical no-traffic yield based on trafficked area and farm average RTF yields.

Apart from the potential yield increase, the controlled traffic system may result in increased silage quality, mainly from increased crude protein contents from a more desirable botanical composition with larger clover contents. To evaluate the potential increase in profit from a more protein rich silage, the model includes adjusted feed rations according to the assumed outcome of CTF in grass clover leys. The value of differences in silage quality is analysed with four alternative feed rations based on three potential scenario outcomes from CTF in terms of silage quality.

Scenario 1 The CTF system reduces the trafficked field area by 50–55% compared to the RTF system depending on the selection of CTF machinery system. This gives a potential yield increase from 10.8 to 11.6% from Eq. (7.2). Scenario one assumes that CTF will provide a yield increase, but that silage quality will remain constant in comparison to RTF.

Scenario 2 In this scenario we assume an increased silage quality from a 10% increase in clover contents and increased yield according to Eq. (7.2). A more desirable botanical composition with an increase of 10% in clover content is assumed to result in a 25% increase of crude protein in the silage.

Scenario 3 If CTF can result in quicker regrowth for the second and third grass cut this may provide an opportunity to advance the second and third cut, which can result in higher energy silage. Scenario three assumes similar herbage yields from RTF and CTF, but increased crude protein and energy contents from controlled traffic. Energy increases from 10.5 to 11.2 MJ kg DM $^{-1}$ and crude protein increase by 25%.

Four different feed rations are presented in Table 7.3 where feed ration one (FR 1) is made for the RTF system and feed rations two to four (FR 2- FR 4) are based on the CTF scenarios presented above.

Table 7.3 Feed rations used in the optimization model

Feed rations and nutritional content								
Feed rations	Silage price = 0 (production cost)			Grain (barley) price = 143 € t^{-1}		Protein concentrate price = 400 € t^{-1}		Price/cow (€)
	MJ kg DM $^{-1}$	Cp as g kg DM $^{-1}$	Feed intake (relative)	MJ kg DM $^{-1}$	Cp as g kg DM $^{-1}$	MJ kg DM $^{-1}$	Cp as g kg DM $^{-1}$	
FR 1	10.5	120	100	13.1	123	13.6	305	836
FR 2	10.5	120	110	13.1	123	13.6	305	791
FR 3	10.4	150	110	13.1	123	13.6	305	770
FR 4	11.2	150	120	13.1	123	13.6	305	700

MJ megajoules, *DM* dry matter, *Cp* crude protein

7.4.2 Optimization Model

The optimization model is based on a maximization problem, formulated as an objective function.

$$\max \pi = IRTF\, C_{RTF} + ICTF\, C_{CTF} + \sum_{j=1}^{J} Gm_j x_j - S_{RTF} c_{RTF} - \sum_{f=1}^{F} S_{CTF_f} c_{CTF_f} + \sum_{i=1}^{I} L_i \left(Gm_i - FR_i \right)$$

$$IRTF, ICTF, x_j, S_{RTF}, S_{CTF_f}, L_i, FR_i$$

Where,

x_j	Quantity hectares of crop j
Gm_j	Gross margin for one hectare of crop j
S_{RTF}	Quantity, hectares of RTF silage
S_{CTF_f}	Quantity, hectares of CTF silage for CTF silage f for $\forall f = 1\ldots3$
C_{RTF}	Production cost for one hectare of RTF grass ley
C_{CTF_f}	Production cost for one hectare of CTF grass ley f for $\forall f = 1\ldots3$
L_i	Quantity of Livestock (dairy cows) i, where i defines feed ration $i = f$
Gm_i	Gross margin from one unit of L_i
FR_i	Feed cost for grain and feed concentrate for one unit of L_i 1 = RTF and 2–4 = CTF
$IRTF$	Binary control variable for RTF machinery system (0, 1)
$ICTF$	Binary control variable for CTF machinery system (0, 1)
C_{RTF}	Annual total capital cost for RTF machinery system
C_{CTF}	Annual total capital cost for CTF machinery system

The maximized total farm profit is determined by fixed and variable machinery costs, accumulated gross margins for each crop, production cost for each hectare of silage and finally gross margin and feed costs for the dairy cows.

The crop rotation includes; oats, spring barley, winter wheat, oilseed rape (OSR) and grass ley. The crop rotation is subject to changes in the optimization model. Oats and spring barley are under-sown with grass, hence, the grass ley area defines the oats and spring barley area. Furthermore, the optimization model selects a machinery system based on fixed cost and yield output for CTF and RTF. The machinery systems are defined as integer variables, which consequently implies that only RTF or CTF can be used.

The optimization model is subject to various constraints which define the crop rotation, silage protein and energy production and feed ration balance. These are described further in Alvemar (2014). The comparison of the optimized total farm profits from RTF and the three simulations of CTF constitute scenarios one, two and three in the results. The three potential outcomes of CTF are compared for each machinery system, 8, 9 and 12 m CTF as previously presented.

7.5 Results

The results are presented as the total farm profits comparing the three alternative scenarios for CTF in grass and the three proposed machinery systems (8, 9 and 12 m CTF) with the RTF scenario. The summary of results in Fig. 7.5 reveals that in all three scenarios the 8 m CTF system gives the largest total farm profits closely followed by the 9 m CTF system. The CTF machinery systems are profitable in all cases with significant increases in total farm profits.

In scenario one, a yield increase is assumed, using Eq. (7.2) and the trafficked field area for each machinery system. From the results it is evident that the combination of a 10.8% yield increase and a relatively small increase in machinery costs for the 8 m CTF system provides a substantial increase in total farm profits compared to the RTF system. The 9 and 12 m CTF systems create marginally lower total farm profits compared to the 8 m system, but still an increase compared to the RTF system.

Scenario two,(increased herbage yield and silage quality) proves to create the highest total farm profits for all CTF machinery systems. The 8 m CTF system gives a 90% increase in total farm profit, closely followed by the 9 m CTF system at 85% total farm profit increase. Again the 12 m CTF system provides the lowest total farm profits of the CTF systems, however, for scenario two the 12 m system shows a 53% increase in total farm profits.

The third scenario is based on the assumption that CTF enables an early harvest time, thus providing high energy and protein silage, but without any yield increase.

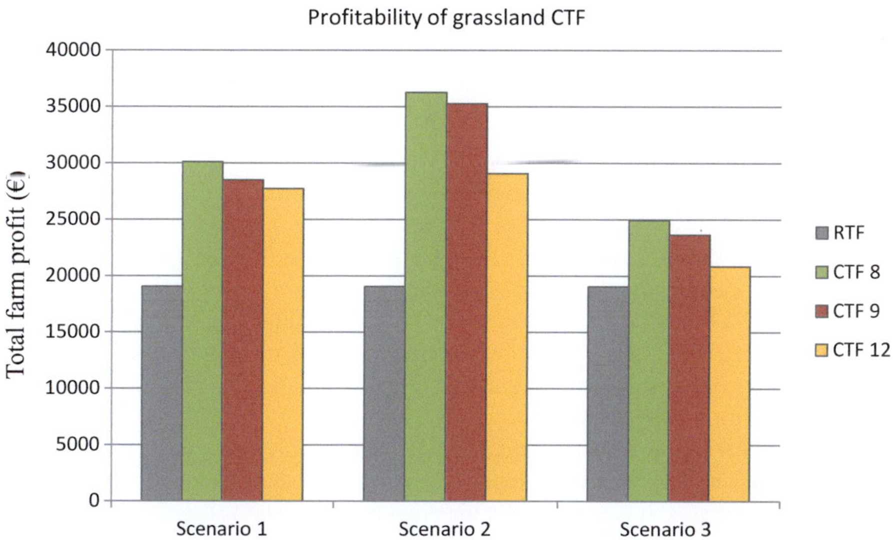

Fig. 7.5 Total farm profits for each machinery system and CTF scenarios based on the 300 cows herd

The 8 and 9 m CTF systems result in 31 and 24 total farm profit increase, respectively. For this scenario the 12 m CTF system creates the overall smallest increase in total farm profits at 9%.

The maximization model is used to generate the maximum total farm profits subject to crop rotation, feed ration and machinery system. The results for each scenario and machinery system are presented in Table 7.4.

The model alters the crop rotation and feed ration based on the machinery system and the silage yield associated with each machinery system. The RTF machinery system is used as a relative measure to compare the CTF systems in each scenario. For scenarios one and two the CTF yield increases as the trafficked field area decreases from the 8, 9 and 12 m CTF systems. For scenario one the major factor controlling profit is reduced grass area in favour of cash crops; in this case winter wheat. It is assumed that grass leys managed with a controlled traffic system are harvested for 4 years rather than 3 years in an RTF system. This assumption is in accordance with experience from Danish farmers practising CTF, and Hansen (1996) showed that the yield increase from non-compacted plots originates from the last years of ley and a botanical composition with larger clover content in year four. This assumption reduces the required land area for barley and oats with under-sown grass by 25%, which may be re-allocated to growing winter wheat. In scenario one, feed ration two (FR 2) with higher silage quantity at lower feeding costs per cow is available. However, the optimal solution uses feed ration one (FR 1), which indicates that more silage of similar quality does not automatically imply that this is the best way to allocate the benefits of CTF. Hence, the increase in total farm profits from CTF for scenario one stems from a reallocation of land resources in the crop rotation with decreased grassland in favour of winter wheat production.

Scenario two produces the highest total farm profits. In this case the profit increase is based on a mix of increased winter wheat production and reduced feeding costs. The land available for increased winter wheat production is mainly a product of the assumption that CTF allows for 4-year grass leys. The increase in silage yield is used in feed ration three (FR 3), which provides a significant reduction in feeding costs per cow (8%).

If the CTF system does not provide a yield increase, but high silage quality in terms of energy and protein, this may also create increased farm profits. For scenario three, the 8, 9 and 12 m CTF systems result in higher total farm profits than the RTF system. For scenarios one and two the grassland areas are reduced when using CTF enables increases in cash crop production. Moreover, in scenario three, grassland areas are increased to meet the increased silage quantity in feed ration four (FR 4), which leaves reduced potential for cash crops to contribute to the total farm profit.

In general for all scenarios, a yield or silage quality increase from CTF is a benefit that contributes to the overall profitability of the system. However, when machinery systems are compared it is evident that the crucial component is to minimize the cost associated with introducing the CTF system. The fixed machinery costs derived from the cost of capital is presented in Table 7.5. As the CTF systems require investments in machinery the fixed machinery costs rise with the size of the module width.

Table 7.4 Summary of optimization results

Scenario. 1	**Grass (ha)**	**W-Wheat (ha)**	**S-Barley (ha)**	**OSR (ha)**	**Oats (ha)**	**Feed ration**	**Variable MC € ha^{-1}**	**Farm profit (€)**	**Farm profits relative to RTF (RTF = 100)**
RTF	161.5	50.5	26.5	35	26.5	Fr 1	383	19,050	100
CTF 8 m	145.8	82.8	18.2	35	18.2	Fr 1	369	30,108	158
CTF 9 m	145.6	83.0	18.2	35	18.2	Fr 1	371	28,502	150
CTF 12 m	144.6	84.2	18.1	35	18.1	Fr 1	325	27,739	146
Scenario. 2	**Grass**	**W-wheat**	**S-barley**	**OSR**	**Oats**	**Fr**	**Variable MC € ha^{-1}**	**Farm profit (€)**	
CTF 8	158.7	66.7	19.8	35	19.8	Fr 3	373	36,217	190
CTF 9	158.5	66.9	19.8	35	19.8	Fr 3	371	35,276	185
CTF 12	157.5	68.1	19.7	35	19.7	Fr 3	321	29,094	153
Scenario. 3	**Grass**	**W-wheat**	**S-barley**	**OSR**	**Oats**	**FR**	**Variable MC € ha^{-1}**	**Farm profit (€)**	
CTF 8	186.6	31.8	23.3	35	23.3	Fr 4	378	24,928	131
CTF 9	186.6	31.8	23.3	35	23.3	Fr 4	374	23,627	124
CTF 12	186.6	31.8	23.3	35	23.3	Fr 4	315	20,851	109

Table 7.5 Fixed machinery costs for RTF and CTF machinery systems

Fixed machinery costs				
	RTF	CTF 8	CTF 9	CTF 12
Capital cost (€)	95,330	101,429	102,929	111,534
Cost per ha. (€)	318	338	343	372

Note: *MC* machinery cost

Table 7.6 Break-even yields for CTF

CTF yield increase break-even analysis			
	Scenario 1 (%)	Scenario 2 (%)	Scenario 3[a]
CTF 8	4.0	2.4	N/A
CTF 9	4.8	2.9	N/A
CFT 12	5.5	6.4	N/A

[a]Scenario 3 is not based on yield increase

The CTF systems increase the fixed machinery costs, but from a cost per hectare perspective the increase in fixed costs is relatively low. Given the potential benefits of CTF, an increase of fixed machinery costs in accordance with the 8 m system, €20 ha^{-1} must be regarded as a limited expenditure.

To emphasize the potential of the CTF system, a break-even analysis is performed. Table 7.6 displays the yield increase required to perform break-even for scenarios one and two. For scenario two, 8 m CTF, merely a 2.4% yield increase will justify the investment in CTF compatible machinery. For the 12 m CTF systems, the break-even yield increase is higher at 5.5% and 6.5% for scenarios one and two, respectively.

Scenario three is not available for the yield increase break-even analysis because this scenario implies that CTF does not provide a yield increase.

To summarize, the results show that CTF in grassland is profitable for all three machinery systems and the three yield and quality scenarios. Furthermore, the machinery costs associated with the machinery system have a substantial impact on the profitability and the break-even yield increase level.

Other studies have also found increased profitability from CTF systems. Stewart et al. (1998) who focus on the effects of machinery traffic in grass production found a 19% gross margin increase for zero and reduced ground pressure systems. However, Stewart et al. (1998) did not consider the value of increased silage yield in a feed ration or alternative land use perspectives. Other studies evaluating the economics of CTF in arable rotations have found increased profits from CTF. Kingwell and Fuchsbichler (2011) did a study in Australia evaluating the benefits of CTF and showed an increase in farm profit of 51% to 67%. Alvemar and Johansson (2013) compared three CTF machinery systems for arable rotations and found a gross margin increase of € 73 per hectare from CTF.

7.6 Discussion

Although the results presented in this study show significant economic benefits of the CTF system, there are still some uncertainties to be considered when converting to controlled traffic. The aspects to consider are: magnitude of machinery investments associated with CTF conversion, expected yield increase and the value in terms of feed rations of any yield increase.

Trafficked Area Investment Ration

The results presented emphasize the importance of keeping investments low when converting to CTF, and that a major reduction in trafficked field area by a minor investment is more important than achieving the minimum trafficked area if this requires major machinery investments. An approach to evaluate the technical returns on invested capital in terms of CTF is a tracked area reduction and investment ratio, presented in Table 7.7.

The tracked area investment ratio shows clearly that the reduction in trafficked field area from the 8 and 9 m CTF systems requires a relatively small investment for each percentage of reduction, whereas the 12 m CTF system induces a fairly large investment at € 5000 for a one percentage reduction in trafficked area. One option to reduce the investment cost may be to use less expensive auto steering and positioning systems.

The results from Hansen (1996) showed a yield increase of 22% from non-trafficked grass-clover leys compared to 100% trafficked plots, therefore, one can assume that a reduction in trafficked field area by 1% should increase the yield by 0.2%.

If CTF results in a yield increase for grass silage, clear benefits are obtained. However, it is not until that yield increase is used in either a feed ration or by changes in the crop rotation that a true monetary value can be presented. The possibility to reduce the grassland area in favour of cash crops in a crop rotation is one opportunity to use the benefits of CTF. Moreover, increased silage yields can be used to reduce feed costs by changing to a more silage dominated feed ration.

On the other hand, the specific conditions for each farm as associated with spatial variation mean that CTF might not provide the same yield increase for all fields, and all fields may not have similar potential for CTF. Some geographical areas may offer small field sizes with irregular field shapes, whereas others have large square fields. There are tools that can be used for predicting CTF tracked area

Table 7.7 Summary of tracked area and investment ratio

Tracked field area and investment ration				
	RTF	CTF-8	CTF-9	CTF-12
Tracked field area	72%	21%	20.5%	17%
CTF investment (€)		54,737	72,105	275,579
Expected yield kg ha^{-1}	8000	8865	8874	8933
Tracked area/investment[a]		1073	1400	5011

[a]Investment cost (€) per unit (%) reduction in trafficked field area

and defining tramlines based on field shape. Tracking direction optimization tools that are readily available can help farmers set up their tramline and CTF systems. Such tools may also aid farmers when using contractors e.g. for slurry spreading to ensure that all field traffic is confined to the tramlines and that tramlines can be communicated in a digital format.

One important part of converting to CTF is the cost of auto steering equipment, future systems may reduce costs associated with GPS and auto guidance systems. We have already noted a decrease in prices for tractor auto guidance systems and potentially the next step will be an improvement in stability and accuracy from geostationary correction signal satellites to replace the need for RTK correction signals. At the moment Trimble offers a correction signal, RTX4 with some potential, however start-up time is 15 min to reach 4-cm accuracy. John Deere has just launched their SF3 correction signal which may have some desirable features in terms of CTF. But for now RTK is the only system that is truly reliable when it comes to quick start up times and year to year accuracy, which is required for CTF.

7.7 Conclusions

To conclude, the profitability of adoption of controlled traffic farming in grass silage is determined by machinery costs, herbage yield and silage quality. Even though there is substantial evidence in previous research that machinery traffic in many cases reduces herbage yield, the introduction of CTF systems require attention to detail in terms of machinery costs.

From a farmer's perspective, the potential profitability of converting to CTF is determined by the existing machinery system and the required investment for a CTF conversion. Moreover, the on-farm profitability potential will be determined by the site-specific conditions in terms of yield response from CTF, opportunities to produce other cash crops, and the knowledge and the involvement required in setting up and maintaining the CTF system.

Based on the specific conditions assumed in this study, CTF in silage production increases the total farm profits in a range from 9 to 90% depending on the machinery system and yield and quality outcome of CTF. This implies a profit of 6 to 57 € per hectares.

Acknowledgements This project is partly funded by Stiftelsen Alfa Laval AB:s (AB Separators) fond through KSLA – The Royal Swedish Academy of Agriculture and Forestry.

References

AHDB – Agriculture and Horticulture Development Board (2016) www.ahdb.org.uk

Alvemar H (2014) Controlled traffic for grass silage production - an economic evaluation for dairy farmers. D-level thesis, Swedish University of Agricultural Sciences, Uppsala

Alvemar H, Johansson C (2013) CTF ur ett ekonomisk perspektiv: en analys för storskaliga växtodlingsföretag (Bachelor thesis). Swedish University of Agricultural Sciences, Uppsala

Arndt W, Rose C (1966) Traffic compaction of soil and tillage requirements. J Agric Eng Res 11:170–187

Chamen WC (2011) The effects of low and controlled traffic systems on soil physical properties, yields and the profitability of cereal crops on a randge of soil types. PhD thesis, Cranfield University, Cranfield

Chamen WCT, Longstaff DJ (1995) Traffic and tillage effects on soil conditions and crop growth on a swelling clay soil. Soil Use Manag 11:168–176

Chamen WCT, Watts CW, Leede PR, Longstaff DJ (1992) Assessment of a wide span vehicle (gantry), and soil and cereal crop responses to its use in a zero traffic regime. Soil Tillage Res 24:359–380

Chyba J, Kroulík M, Krištof K, Misiewicz PA, Chaney K et al (2014) Influence of soil compaction by farm machinery and livestock on water infiltration rate on grassland. Agron Res 12:59–64

DiaryCo – Average Herd Size 2016–09–05. http://dairy.ahdb.org.uk/resources-library/market-information/farming-data/average-herd-size/#.V83GuldqOpO

Dickson JW, Ritchie RM (1996) Zero and reduced ground pressure traffic systems in an arable rotation 2. Soil and crop responses. Soil Tillage Res 38(1–2):89–113

Douglas JT, Crawford CE (1991) Wheel-induced soil compaction effects on ryegrass production and nitrogen uptake. Grass Forage Sci:405–416

Douglas JT, Campbell DJ, Crawford CE (1992) Soil and crop responses to conventional, reduced ground pressure and zero traffic systems for grass silage production. Soil Tillage Res 24:421–439

EEA, European Environmental Agency (2015) www.eea.europa.eu

Elonen P (1986) Reports from the Division of Soil Management: Jordpackning – ett problem i Finsk åkerodling. Rapporter från jordbearbetningsavdelningen No. 71. Department of Soil Sciences, Uppsala

Frame J (1982) The effect of wheel tracking on red clover (Trifolium pratense) swards. In: Efficient grassland farming. British Grassland Society, Reading, pp 313–315

Frost JP (1988) Effects on Crop Yields of Machinery Traffic and Soil Loosening Part 1. Effects on Grass Yield of Traffic Frequency and Date of Looseing. J Agric Eng Res 39:301–312

Håkansson I, Reeder RC (1994) Subsoil compaction by vehicles with high axle load—extent, persistence and crop response. Soil Tillage Res 29:277–304

Håkansson I, McAfee M, Gunnarsson S (1990) Verkan av körning med traktor och vagn vid vallskörd. Resultat från 24 försöksplatser. (No. 78). Department of Soil Sciences, Uppsala

Hansen S (1996) Effects of manure treatment and soil compaction on plant production of a dairy farm system converting to organic farming practice. Agric Ecosyst Environ 56:173–186

Husdjur i juni 2015, 2016–09–15. http://www.jordbruksverket.se/webdav/files/SJV/Amnesomraden/Statistik%20fakta/Husdjur/JO20/JO20SM1502/JO20SM1502_tabeller3.htm

Jorajuria D, Draghi L, Aragon A (1997) The effect of vehicle weight on the distribution of compation with depth and the yield of Lolium/Trifolium grassland. Soil Tillage Res 41:1–12

Jørgensen RN, Green O, Swain KC, Kristensen R, Bochtis D, Sørensen CG (2009) Impact on clover-grass yield casued by different traffic intensities. International Agricultural Engineering Conference, Bangkok

Keller T, Arvidsson J (2004) Technical solutions to reduce the risk of subsoil compaction: effects of dual wheels, tandem wheels and tyre inflation pressure on stress propagation in soil. Soil Tillage Res 79:191–205

Kingwell R, Fuchsbichler A (2011) The whole-farm benefits of controlled traffic farming: an Australian appraisal. Agric Syst 104:513–521
Kroulik M, Paula M, Chyba J, White D (2014) Field traffic intensity for forage harvesting in the UK. HAU Project Report No. 099, Harper Adams University, Newport
Lingvall P, Arvidsson H (2005) Harvesting silage with two types of silage trailer (feed rotor with knives and precision chop). In: Proceedings XIVth Int. Silage Conf. Belf. North. Irel. July 2005, p 192
Lingvall P, Knicky M (2008) Precision chopping or rotor cutting and its influence on enisling capacity and silage fermentation. Grassl Sci Europe 13:666–668
McDonald P, Edwards RA, Greenhalgh JFD, Morgan CA, Sinclair LA, Wilkinson RG (2011) Animal nutrition, 7th edn. Pearson Education Limited, Harlow
Mean Precipitation 2016–09–05. http://www.eea.europa.eu/data-and-maps/indicators/european-precipitation-1/assessment
Nadeau E, Arnesson A, Hallin O (2014) Vallskördeteknikens och tillsatsmedlens betydelse för ensilagets kvalitet [Forage harvesting technique and additives – effects on silage quality] (No. 37). Swedish University of Agricultural Sciences, Department of Animal Environment and Health Section of Production Systems
Pedersen HH, Novak G (2013) Beregning af overkørt areal. AgroTech
Raper RL (2005) Agricultural traffic impacts on soil. J Terramech 42:259–280
Rasmussen KJ, Møller E (1981) Genvækst efter fortørring af græsmarksafgrøder. II Jordpakning i forbindelse med høst og transport Tidskr Planteavl 85:59–71
SJV – Swedish board of Agriculture (2016) www.jordbruksverket.se
Søgaard HT, Sørensen CG (2004) A Model for Optimal Selection of Machinery Sizes within the Farm Machinery System. Biosyst Eng 89:13–28
Stewart LED, Copland TA, Dickson JW, Douglas JT (1998) Economic Evaluation of Traffic Systems for Arable and Grass Crops on an Imperfectly Drained Soil in Scotland. J Sustain Agric 12:41–56
Tullberg J, Yule D, McGarry D (2003) "On track" to sustainable cropping systems in Australia. In: Proceedings international soil tillage research organisation. 16th triennial conference, Brisbane, 13–18 July 1271–1285
Webb B, Blackwell P, Riethmuller G, Lemon L (2004) Tramline farming systems: technical manual bulletin 4607

Chapter 8
Robotic Seeding: Economic Perspectives

Søren Marcus Pedersen, Spyros Fountas, Claus G. Sørensen, Frits K. Van Evert, and B. Simon Blackmore

Abstract Agricultural robotics has received attention for approximately 20 years, but today there are only a few examples of the application of robots in agricultural practice. The lack of uptake may be (at least partly) because in many cases there is either no compelling economic benefit, or there is a benefit but it is not recognized. The aim of this chapter is to quantify the economic benefits from the application of agricultural robots under a specific condition where such a benefit is assumed to exist, namely the case of early seeding and re-seeding in sugar beet. With some predefined assumptions with regard to speed, capacity and seed mapping, we found that among these two technical systems both early seeding with a small robot and re-seeding using a robot for a smaller part of the field appear to be financially viable solutions in sugar beet production.

Keywords Robotics • UAV • Seeding • Profitability

S.M. Pedersen (✉)
Department of Food and Resource Economics, Faculty of Science, University of Copenhagen, Frederiksberg C, Denmark
e-mail: marcus@ifro.ku.dk

S. Fountas
Agricultural University of Athens, Athens, Greece
e-mail: sfountas@aua.gr

C.G. Sørensen
Department of Engineering, University of Aarhus, Aarhus, Denmark
e-mail: claus.soerensen@eng.au.dk

F.K. Van Evert
Wageningen University & Research, Wageningen, The Netherlands
e-mail: frits.vanevert@wur.nl

B.S. Blackmore
Harper Adams University, Newport, Shropshire, UK
e-mail: simon.blackmore@harper-adams.ac.uk

S.M. Pedersen, K.M. Lind (eds.), *Precision Agriculture: Technology and Economic Perspectives*, Progress in Precision Agriculture,
https://doi.org/10.1007/978-3-319-68715-5_8

8.1 Introduction

Agricultural robots have become a recent trend in agriculture with many research and commercial applications already in place, which has caused an increasing awareness and initiated development of agricultural robots around the world, e.g. STOA, EU Scientific Foresight Study (2016). Agricultural robots have significant advantages in terms of flexibility, adaptability and environmental benefits compared to traditional agricultural machinery and represent a great asset for advanced precision agriculture. Many advances in machine vision with imaging and spectral cameras for detecting weeds, pests and diseases, nutrient deficiencies in soil or plant leaves have provided weighty applications to be mounted on autonomous vehicles for scouting soil and crops and for further downstream decision support and control. The robotic community has also made momentous improvements and advancements in terms of using a widespread operating system, the Robotic Operating System (ROS), as well as applying robust platforms and navigation algorithms.

In relation to agricultural robots, the principal systems characteristics of robots are the light weight, small size and energetic autonomy (Fountas et al. 2007). Light weight means that the vehicle requires lower propulsion energy and induces less soil compaction while at the same time the vehicle must be small for safety reasons, for achieving greater precision during execution of tasks and having more manoeuvrability within the field to minimize turning time lags. The mechanical design of the prototypes depends on the main tasks that the vehicle has to carry out. To achieve the maximum manoeuvrability, which is very important for autonomous vehicles, four-wheel drive and steering (4WD/4WS) is commonly used (i.e. BoniRob robot at Ruckelshausen et al. 2009; HortiBot robot at Sørensen et al. 2007; AgRover robot at Tu 2013). To be able to work in different types of crops and for maximum flexibility, many agricultural robotic prototypes have variable track width and height configuration (Bonirob, Hortibot and AgRover robots). Problems with this type of approach have been reported because a vehicle's centre of gravity, especially when the height increases, will make prototypes unstable at slopes. To reduce stability problems on slopes, HortiBot has a very low centre of gravity to be able to work on slopes up to 40°. AgRover robot has a self-levelling pneumatic system for maintaining the platform flat and stable. In Appendix there is a list of current agricultural robots for arable farming and orchards that are under development or commercially available.

In general terms, energy efficiency and autonomy are the major issues for agricultural robots today, and robots should be constructed to target energy efficiency with lightweight robots using light but strong materials like carbon fibre and aluminium or by using decentralised modern low power controllers and electronics where possible, instead of a single high speed central processor. In terms of interoperability, current solutions have limited compliance to standards or limited access to the control system implementation which hampers their integration in robot–robot or human–robot teams. Finally, one of the main challenges regarding unmanned systems relates to safety. Even if the human is remotely controlling the robot, a malfunction could lead to problematic situations. A robust solution is to shut down the system following any component failure (BoniRob). However, this is undesirable because the whole process must stop during recovery. Interoperability between agricultural machines is still very limited and cooperation between agricultural robots is at the research level.

8.2 Economic Performance

For most operations, agricultural robots have so far not been affordable because of high investment and maintenance costs compared to cheap labour in many countries. However, in the years to come, the technology will become cheaper and labour costs will become more expensive. One example from horticulture of an economically viable system is a strawberry-harvesting robot that moves on rails in a greenhouse with a camera system to image the berries and judge which one to pick according to colour of the fruit. It was launched in 2013 at a cost of 50,000 US$ and is expected to be cost-equivalent to human labour in Japan when used at the large scale. In the open field, only a few weeding systems in lettuce and other high value crops are offered commercially so far.

Other studies based on pre-commercial systems in arable farming have also focused attention on the feasibility of agricultural robots on grass both moving and weeding with promising perspectives (Pedersen et al. 2006). Sørensen et al. 2007 assessed the feasibility of a plant nursing robot for weeding operations and found that profitability gains ranging from 20 to 50% are achievable through targeted applications.

A few commercial companies like Vitirover has already provided guidelines about the investment cost of a small rover that removes weed in viticulture. The capacity is about 0.4 hectare per unit and the cost is 5000 EUR. The company expect a depreciation time of about 6 years for their robotic solution, which is similar to 833 EUR ha^{-1}, which is about the same as conventional weeding practices in vineyards 600–1200 EUR ha^{-1}depending on weeding techniques. (www.vitirover.com).

However, so far little attention has been given to quantifying the profitability of operations in arable farming. Therefore, the overall aim of this chapter is to quantify the economic and environmental benefits from the application of agricultural robots under specific conditions and constraints. The objectives will be to explore these benefits for crop establishment (seeding) in a selected crop with a large yield and turnover, i.e. sugar beet.

This assessment is based on a comparison of gross margins of conventional cultivation practices in sugar beet and new seeding systems. Initial investment costs and ongoing financial costs and benefits are quantified on a yearly basis and applied to specified mobile robotic systems in different applications. It is based on the assumption of models building on the common features among mobile robots in relation to labour costs, speed and capacity, and expected lifetime of the systems.

8.3 Robotic System for Crop Seeding

Soil and seeding practices are essential for preparing a proper seed and rooting bed for crop production and has major effect on the productivity and environmental footprint of agriculture. At a global scale, poor quality of seed establishment because of improper tillage and seeding operations can cause extensive yield losses and poor use of external inputs such as energy, soil, nutrients, water and pesticides. Stagnating cereal yields in Northern Europe calls for a major rethinking of tillage and seed establishment involving the use and development of intelligent tillage and seeding methods to be closely adapted to the specific conditions on the farm and within the field as well as by using the newest technologies in terms of sensors and autonomous systems.

The above mentioned factors are the key factors for setting up the following scenarios encompassing robotic systems for seeding that can improve conventional methods through, for example, precision seeding as opposed to ploughing and broad seeding; reseeding places where seeds have not emerged could be reseeded precisely with a machine vision system to recognize the missing plants over an entire field. The system's use of lighter machinery and more intelligent tools will reduce the negative impact of machinery on soil compaction.

A **Robotic platform** for high accuracy seeding is characterised by having different features. The platform is ultralight and provides minimum compaction. It moves at a slow speed of 3 km/h and requires as little energy as possible given the size of the vehicle. Although it may not be as energy efficient as large tractors because of its small scale configuration, it has relatively low fuel costs because operations are precise with minimal overlaps. The robot, with its low weight and autonomy, has an extended window for field readiness (execute operations at the optimal time for crop development and not when the soil can sustain heavy machinery). It is possible to optimise logistics (e.g. load of seed and fertiliser synchronised with track and field length). It is scalable implying that the robot can be fitted to both large and small fields not suited for large machinery, meaning that that small and medium farms can benefit from economies of scale. With these characteristics, it provides significant increases in operational flexibility compared to larger machines.

Accurate crop seeding with robots has some further characteristics.

The system has the ability to place the seed accurately within 2 cm of a known point (RTK GNSS), to reduce intra crop competition and eliminate the need for thinning.

For instance, orientation of seed has shown that maize seeds, and hence plants can be aligned so that there is no intra crop competition for sunlight.

For these systems, a zero draft force powered punch planter, vertical axis rotary cultivator or vertical tube planter can be used to place each seed positively. The ability to change seed spacing according to soil type, moisture regime may be used as a mitigation factor when planting high value crops to achieve a more uniform size. (e.g. less onion seed in drier areas and more onion seed in wetter areas to get a larger proportion of saleable crop).

Seeding to specific depths allows seeds to be put deeper into the soil to ensure it comes into contact with soil moisture, which usually increases with soil depth. In addition, starch and water gels can be used to help germination where inherent soil moisture is insufficient for germination or where the soil to seed surface may not be enough. The use of gels will reduce cultivation caused by increased contact between the seed surface and the soil structure as well as having intrinsic water availability within the gel to trigger germination. Another characteristic incorporated into the starch gel is the timely release of nutrients and or chemicals to promote or retard pests, diseases and growth. Moreover, proximity fertilisation can be achieved by placing fertiliser at an even distance from the seeding position to ensure maximum uptake of nutrients and minimise leaching to groundwater.

A permanent planting position allows crop residues to be kept from the seed position and the same seeding map can be used each year. Conversely, the seeding map can be offset to ensure all seeds are put into new soil positions. As each seed is handled separately, it would be possible to have as many varieties or species of seed as there are seeding positions. Microclimates occur in most fields (shading under trees, different soil types, etc.) and could be seeded with specific varieties suitable for each situation. Seed rates could be also varied.

Reseeding is the ability to go back into a planted field and re-seed places where germination or emergence has not occurred or plants have been damaged by rabbits, flooding, etc. Reseeding is not usually available because tractors would do too much damage to the existing crop. Transplanting could be used if areas of crop were damaged, but the existing crop had matured too far to allow reseeding.

Seed mapping can be used to record the position of each seed that goes into the ground. This information can be used to predict the probability of location of each crop plant for subsequent operations like mechanical weeding or individual plant care operations.

8.4 Economic Assessment of Early Seeding and Re-seeding with Robots in Sugar Beet

In this example, we have assessed the potential gross margin of robotic seeding in sugar beet. In principle, we have compared the gross margin of conventional production of sugar beet with two different technical scenarios:

Early seeding
Site-specific re-seeding of areas with poor seed emergence

For simplicity reasons, we have assumed that the capacity is 500 ha of arable farm land and that all scenarios are focusing on sugar beet production under Danish farm conditions. Conventional treatment is regarded as the base scenario and yields, prices, variable and fixed costs are based on budget data for year 2015 from Landbrugsinfo, A national Danish farm portal administered by the advisory centre

SEGES. The soil type for all scenarios is regarded as an average soil (JB 5-6) which is a good clay soil with relatively good yield levels. Yield in conventional sugar beet production is assumed to be 60,000 kg per ha. with an average price of 0.047 € per kg sugar beet which is equivalent to 3152 € per ha. Waste production is 24,000 kg per ha with a total value of 32 € per ha. In this basic or conventional scenario, the gross margin after payment of variable cost (including machinery and labour costs) is 1238 € per ha. With this base scenario as a reference scenario we attempted to assess the marginal change in cost and additional benefits if a farm manager decided to implement one of the two new systems (early-seeding and re-seeding) and compare it with the conventional reference scenario.

For both scenarios, we have assumed that investments in technology such as UAV systems, early seeding and re-seeding systems are expected to depreciate over a period of 5 years.

For the individual technical systems, we have made the following assumptions as described below.

8.5 Early Seeding

A study in Australia by D.R Coventry et al. (1993) on sowing time indicates that yield decreases with delays in sowing time. In a 2-year experiment there was a loss of 200–250 kg grain ha in 1985 for each week's delay in sowing time and similar 50–110 kg grain ha^{-1} loss per week in 1987 in wheat.

A study by R. K. Scott et al. (1973) indicates that yield decreases in sugar beet by 0.4 tons per ha for each week it was delayed from early April until the beginning of May. For later sowings the yield loss was greater. We have assumed in this study that the relative difference between sowing in the conventional scenario and early seeding scenario is about 4 weeks, which is equivalent to a difference in yield at 1600 kg or 2.67% compared with conventional treatment because of early seeding and better seed establishing conditions.

In a study by Smit (1990) in The Netherlands it was determined that every day earlier for sowing gives an additional 100 kg sugar per hectare.

Seeding is carried out with a low weight autonomous vehicle and a small adapted seeding unit.

The Zeus robot is a low cost system. This prototype robot (Fig. 8.1) that was constructed at Harper Adams University, UK, with the cooperation of the Agricultural University of Athens will be used in the scenario calculations. It has two-wheel drive and two-wheel steering, and it uses petrol as a power source. It has a petrol engine ATV coupled with a centrifugal CVT transmission that allows forward speed control by the throttle cable and cable operated brakes on both front and rear. The compact design of the ATV is ideally suited for operations within sugar beet fields. The forward speed of the platform has been reduced by changing the original sprockets and by adding to more additional sprockets to the transmission system. The robot operates with the Robotic Operating System (ROS) software architecture.

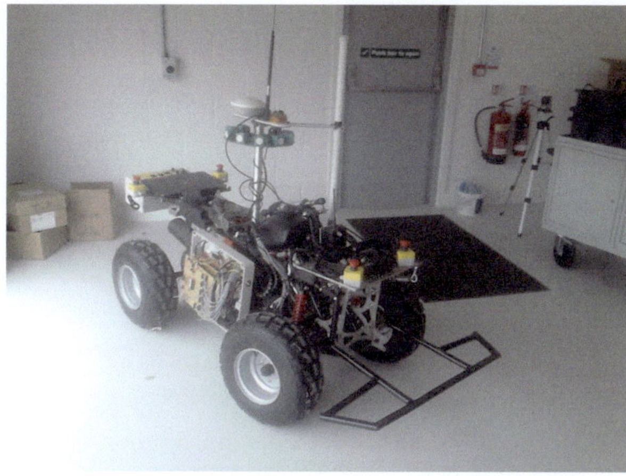

Fig. 8.1 Zeus robot at the lab

For our assumptions, we will take into consideration the 400 cc Kymco model that has the capability to tow a trailer with a 400 kg of load. In this scenario, such a low weight ground robot could work with a newly designed light-weight seeder at 3 kmh. Such seeders do not exist in the market yet, but they have to be designed to work with light weight ground vehicles in soil conditions where conventional seeders would be too heavy and cause soil compaction. It is assumed that such a seeder will have a width of about 1.5 m, seeding two rows of sugar beet. The fuel consumption is estimated to be 3 l h^{-1} with the seeder. In case, we need a larger seeder, then a tractor-based autonomous vehicle could be the other alternative, such as the Hako tractor, developed in Denmark (Reske-Nielsen et al. 2006).

The seeding unit is assumed to have a width of 1.5 m implies that the robot has to run about 7 km per hectare which is about 2.3 h per ha with a speed of 3 kmh. Fuel consumption is then assumed to be about 6.6 l per ha, which is about 4 € ha^{-1} with a fuelprice of 0.6 € per litre. A small robot (or tractor) will use more fuel per ha than a large tractor. For comparison a conventional tractor uses about 3–4 l per ha for seeding.

At this early time of the growing season, it will be difficult for heavy vehicles to operate because of wet soils. In addition, as the robotic system is autonomous the farmer will save labour time and the cost of conventional seeding will be replaced by the robotic system as the robot will carry out seeding on the entire field.

Initial investment costs of the robotic vehicle is assumed to be 25,000 € and the seeding unit is assumed to cost 20,000 €. It is further assumed that two units are needed for early seeding to cover 500 ha. For re-seeding it is assumed that one unit is able to cover the damaged are which will then have to be below 50% of the field. Maintenance of the robot and seeding unit is about 6.5 € ha^{-1} and 3 €/ha for the UAV system. In addition, labour is added at a rate of 27 €/h with an average of 6 min ha^{-1} for crop scouting with the UAV as well as the cost for stitching. Another 2.7 EUR ha^{-1} is added in labour costs for implementing the robot which is about 1300 € in labour cost for the entire field.

8.6 Re-seeding

The re-seeding task has two steps: locate the areas where emergence is so poor that re-seeding must be undertaken, and performing the re-seeding. It is thought that areas with poor germination will be located by processing aerial photographs that are taken with a digital camera mounted on aUAV. When the areas to be re-seeded have been defined, a re-seeding map and route for the seeding robot can be created.

8.7 Creating Re-seeding Maps with UAVs

The UAVs can be either a fixed wing solution or a rotary wing solution, the fixed wing can work at a higher speed with greater capacity, whereas the rotary system is more flexible. There are various options to use. Currently, the price of a professional UAV (Dandrone hexodrone) including a normal digital camera and NDVI camera but without software is about 5300 €. https://dandrone.dk/shop/landbrugsdrone. With three batteries, the pilot is able to use the drone for 3 × 20 min. Hereafter, additional charging is needed. Software is expected to cost about 200–500 € per month. DroneDeploy is a software that can be used to navigate the drone and to processing field maps quickly. The costs of this software for a 1–5-cm pixel resolution vary between 1000 and 5000 UAS per year for the software. With a 1-cm pixel resolution it is about 4000 € per year.

Another less expensive system that is commercially available is the DJI Phantom Drone. It has an action radius of 300 m. It has a maximum speed of 10 meter per second horizontal and use 3S LiPo-batteries, which takes about 1 h to refuel. Each battery is useful for about 10–15 min.

A wing-based drone, like the Trimble UX5 is another option, but currently more expensive; it can be airborne in 45 min with a speed of 80 km per h. The price is about 23,000 €. The AgDrone System™ is another complete solution targeted for the agricultural sector http://www.honeycombcorp.com/products/

In this study, we assume that the system is based on a Dandrone hexodrone including a normal digital camera and NDVI camera and software.

To locate the week spots in the field where seeds have not germinated and to assess the need for re-seeding, crop scouting is made by UAVs that monitor the field during the growing season. It is estimated that the flying time is about 6 min per hectare in a commercial setting. Images made by a camera mounted on the drone are then further processed with stitching and made into a seed map. Based on these seed maps seeding is applied in targeted site-specific doses with a seeder mounted on the robot.

Crop scouting has to be done as timely as possible, implying that a UAV drone is needed in a targeted time span.

Crop scouting with a UAV is assumed to cost 3,6 € ha^{-1} and stitching and processing are expected to cost 8 €/ha. In this respect the UAV is expected to cost about 5300 € with an expected depreciation time of 5 years.

The seeder is used in two scenarios. In the first, it is assumed that the average increase in yield is 5% from re-seeding in patches on the field. In reality, the variation in area that need re-seeding may be relatively high, which the farmer will have to assess each year if it is necessary to re-seed. In some years, much more severe damage may have occurred on the field exceeding the 5% area considerably and in other years it may be 1–2% only of the field that is damaged.

To implement this scenario in practice we installed a low weight robot for seeding and a seeding unit as in the case for the early seeding scenario. It is assumed that the Drones will make the map for showing the non-seeding spots to execute re-seeding.

It is assumed that conventional seeding is still needed because re-seeding is regarded as a method to repair the field without a fullly re-establishing or seeding the entire field. Additional costs for this operation will include 5% extra seeds for re-seeding.

8.8 Results

Table 8.1 below presents the gross margin after the costs of machinery and labour for the three different scenarios (1) Conventional practice (business as usual) (2) Early seeding and (3) re-seeding.

The most profitable system is the early seeding. Although it is assumed that the potential yield increase is only 2.5% compared to conventional seeding, it is expected that the cost of seeding is less because of savings of labour as this operation is now carried out by the seeding-robot. In the re-seeding scenario, it is expected that the yield increases by 5% compared with conventional practice, but in this case the farmer will still have to practice conventional seeding because re-seeding is regarded as a repairing action that requires two seeding operations.

Findings from this study and based on the current assumptions indicate that among these two technical systems both early seeding with an autonomous vehicle and re-seeding appear to be financially viable solutions under the given assumptions.

The gross margin after cost of machinery and labour costs in the reference scenario with conventional sugar beet production is 1238 € per ha.

The most feasible system is early seeding because it includes only the cost of seeding one time without additional seeding as is the case for re-seeding. With early seeding it is possible to increase gross margin with 7.7% from 1238 € ha^{-1} to 1333 € ha^{-1} and with re-seeding it is possible to increase gross margin with 6.5%. From 1238 € ha^{-1} to 1319 € ha^{-1}.

Table 8.1 Gross margins for conventional seeding, early seeding and re-seeding in sugar beet

€	Conventional	Early seeding	Re-seeding
Yield			
Beet	2800	2875	2940
Waste	32	32	32
Transport compensation	320	320	320
Total yield	**3152**	**3227**	**3292**
Variable costs			
Seed	−272	−272	−286
Fertilizer	−108	−108	−108
Phosphorus	−53	−53	−53
Potassium	−120	−120	−120
Herbicides	−200	−200	−200
Fungicides	−28	−28	−28
Other transport	−400	−400	−400
Variable costs	**−1181**	**−1181**	**−1195**
Gross margin	**1971**	**2046**	**2097**
Cost of machinery and labour costs			
Ploughing	−90	−90	−90
Post harrowing	−27	−27	−27
Fertilizer spreader	−19	−19	−19
Seedbed harrowing	−43	−43	−43
Seeding (conventional + re-seeding + crop scouting)	−67	−46	−112
Compaction	−21	−21	−21
Spraying	−107	−107	−107
Inter row cleaning	−53	−53	−53
Harvest	−240	−240	−240
Other operations	−67	−67	−67
Cost of machinery and labours costs	**−733**	**−712**	**−778**
Gross margin after Cost of machinery and labours costs	**1238**	**1333**	**1319**
Percentage change of gross margins compared with conventional		**7.7**	**6.5**

Note: Exchange rate 1 € = 7.5 DKK

8.9 Sensitivity Analysis

In principle, **early seeding** could be executed without robots. Instead of robots, small conventional low weight tractors could be used because these will not compact wet soil as is the case with heavy machinery. However, small machines require high labour costs per unit area because of their to small capacity unless fleets with small machinery units are considered (Sørensen and Bochtis 2010).

In addition to an extra yield benefit from early seeding, a further benefit from low weight robots or small tractors could be reduced cultivation. Without compaction from small machines, there is no need to cultivate the soil. In the light of that, it may be possible to save further operational costs in relation to ploughing (90 € ha^{-1}), post harrowing (27 €), seedbed harrowing (43 € ha^{-1}) and compaction (21 € ha^{-1}). All these cost savings will add up to an additional 181 € ha^{-1} in savings by using low weight vehicles for seeding. However, to reduce cultivation to a minimum it is required that other operations such as spraying, fertilizer spreading and harvesting operations are kept to a minimum to reduce compaction further or executed using small vehicles.

As indicated above, in the **re-seeding** scenario it is assumed that the average increase in yield is 5% which reflects a re-seeded area of 5% of the damaged patches on the field. In reality, the damaged area may vary significantly from year to year. In this regard, a sensitivity analysis is done to depict what effect the difference in percentage of damaged areas in the field has on gross margins, and to assess when a system with autonomous re-seeding becomes profitable compared with conventional seeding depending on the field conditions.

As indicated in Table 8.2, the breakeven point of re-seeding is around 2% of the damaged area. Under the above assumptions it appears that minor damage only to the field may suggest that it is worthwhile to reseed the damaged area with this concept. Sugar beet is regarded as a high value crop and the turnover per hectare is relatively large compared with cereals and oil seed crops.

However, the concept of robotic re-seeding may also be applicable in these crops or in other high value crops.

In this study we have disregarded cost of extra safety precautions. If the robot has to be fully surveyed the profitability will be reduced significantly. Alternatively extra investment like fences, cameras and monitoring of the system while doing other tasks on the farm may be a solution.

Precautions should also be taken with regard to investment. In principle, we have assumed that the cost of a robotic unit will be about 25,000 € and a seeding unit about 20,000 € with a capacity to cover 500 ha of arable land per year. The individual farmer is most unlikely to have 500 ha of sugar beet, but such a system could serve a group of farmers or be based on contracting. The price is assumed to reflect the price of a commercial system that is available in the near future and produced in

Table 8.2 Difference in gross margin of re-seeding in sugar beet compared with conventional seeding (with damaged area from 1–10%) different percentage are of re-seeding

Percent of area that is re-seeded	1	2	3	4	5	6	7	8	9	10
1. Gross margin Re-seeding, €	1218	1243	1269	1294	1319	1345	1370	1395	1420	1446
2. Gross margin Conventional seeding, €	1238	1238	1238	1238	1238	1238	1238	1238	1238	1238
Difference (1–2), € [a]	−20	5	31	56	81	107	132	157	182	208

Note: [a]A positive value is in favour of re-seeding

large quantities and suitable for different seeding operations. The area of sugar beet in 2014 was only 4.5 million ha worldwide in 2014 according to FAO (FAO statistics). If a robotic system is implemented on just 10% of this area it is equivalent to about 900 units. In this respect, the development and commercialization of these systems should have a broader application in different type of cropping systems to be of interest to companies that produce farm machinery.

8.10 Conclusions

In this chapter, we have assessed the potential gross margin of robotic seeding in sugar beet. In principle, we have compared the gross margin of conventional production of sugar beet with two different technical scenarios: early seeding and re-seeding. Findings from this study indicate that among these two technical systems both early seeding with an autonomous vehicle and re-seeding of a small part of the field (5%) appears to be financially viable solutions in sugar beet.

Appendix: Selected Companies That Develop Robots for Different Farm Operations

Company	Website
Naïo Technologies develops and markets robots for agriculture and viticulture	www.naio-technologies.com/en/agricultural-robotics-experts
Precision Makers is a Dutch company The have developed the Greenbot a self-driving machine that has been specially developed for the agricultural and horticultural sectors that perform work tasks that are regularly repeated	http://www.precisionmakers.com/greenbot/
Wall-ye provides autonomous solutions for pruning in vineuard and other crops such as blue berries	http://wall-ye.com/index.html
Blue River is an American company that have produced Smart Agri. Technologies. incl. The LettuceBot – a smart implement that identifies every plant in lettuce, makes a decision based on what it sees, and precisely sprays individual plants	http://www.bluerivert.com/
Deepfield Robotics is a Bosch Start-up Company from Germany. One of their solutions is the BoniRob – a multi-purpose robotic platform for applications in agriculture	https://www.deepfield-robotics.com/index-en.html
SwarmFarm Robotics is an Australian based company. It has developed autonomous crop spraying technology	http://www.swarmfarm.com/

(continued)

Company	Website
SAGA Robotics are developing robots for agriculture to e.g. make accurate yield estimates and early detection of diseases	https://sagarobotics.com/
Agribot is a robot, that autonomously does all the work in orchards and plantations	http://agribot.eu/?lang=en
Dynyam project is a development project that orginates from Oxford and works with development of autonomous agricultural systems	http://www.dyniumrobot.com/
Precision Makers supplies a wide range of solutions for professionals in the agricultural and horticultural sectors: greenbots, robot packages, measuring and control systems, and GPS products	http://dutchpowercompany.com/en/precision-makers/

References

AgDrone System (2016) ™ http://www.honeycombcorp.com/products. Accessed 3 Oct 2016
Coventry DR, Reeves TG, Brooke HD, Cann DK (1993) Influence of genotype, sowing date, and seeding rate on wheat development and yield. Aust J Exp Agric 33(6):751–757
Dandrone (2016) https://dandrone.dk/shop/landbrugsdrone. Accessed 3 Oct 2016
Fountas S, Blackmore BS, Vougioukas S, Tang L, Sørensen CG, Jørgensen R (2007) Decomposition of agricultural tasks into robotic behaviours. Agricultural Engineering International: the CIGR Ejournal. Manuscript PM 07 006. vol IX. October, 2007
Landbrugsinfo (2015) Budgetkalkuler for 2015. www.landbrugsinfo.dk
Pedersen SM, Fountas S, Have H, Blackmore BS (2006) Agricultural robots—system analysis and economic feasibility. Precis Agric 7(4):295–308
Reske-Nielsen A, Mejnertsen A, Andersen NA, Ravn O, Nørremark M, Griepentrog HW (2006) Multilayer controller for outdoor vehicle. In: Proceedings of the Automation Technology for Off-Road Equipment (ATOE), Bonn, Germany, 1–2
Ruckelshausen A, Biber P, Dorna M, Gremmes H, Klose R, Linz A, Weiss U (2009) BoniRob–an autonomous field robot platform for individual plant phenotyping. In: Proceedings of the European conference on precision agriculture (ECPA '09), Wageningen, Netherlands, On CD
Scott RK, English SD, Wood DW, Unsworth MH (1973) The yield of sugar beet in relation to weather and length of growing season. J Agric Sci 81(2):339–347
Smit AL (1990) Teeltvervroeging bij suikerbieten: een verslag van uitgevoerde proeven op het PAGV in de periode 1985 t/m 1988 in het kader van project 56.4.14. Lelystad: PAGV. [Increasing the earliness of sugarbeets: a report of experiments at PAGV during 1985-1988]. Available online at: http://library.wur.nl/WebQuery/clc/514819 and http://edepot.wur.nl/350199
Sørensen CG, Bochtis DD (2010) Conceptual model of fleet management in agriculture. Biosyst Eng 105(1):41–50
Sørensen CG, Nørremark M, Jørgensen RN, Jensen K, Maagaard J, Jensen LA (2007) Hortibot: feasibility study of a plant nursing robot performing weeding operations–Part IV. In: Proceedings of the 2007 ASAE annual meeting, Minneapolis, USA, June, pp 17–19
STOA (2016) Precision agriculture and the future in Farming in Europe. Scientiifc Foresight Study, IP/G/STOA/FWC/2013-1/Lot7/SC5, Dec. 2016, European Parliament
Tu X (2013) Robust navigation control and headland turning optimization of agricultural vehicles. PhD thesis, Iowa State University, USA
Vitirover (2017) www.vitirover.com. Accessed 15 Mar 2017

Chapter 9
Future Perspectives of Farm Management Information Systems

Zisis Tsiropoulos, Giacomo Carli, Erika Pignatti, and Spyros Fountas

Abstract Farm Management Information Systems (FMIS) have evolved from simple record keeping to sophisticated solutions able to capture new trends involving spatial and temporal management, distributed sensors involving interoperability of sensing devices, future internet applications and web services. The FMIS were initially designed to deal with the farmer as the main focus of the system, whereas now data flow from and to the tractor information board, and connections with other pieces of equipment such as precision agriculture devices can be managed through an FMIS. This pathway of evolution has led to the inclusion of a rich set of functionalities and opened up the possibility to improve the cost control of farms. In this chapter, we present the state-of-the-art on these topics depicting the new functionalities included in evolved FMIS and how they can connect the farm to the external context and stakeholders. Then, we delve into the costing functionality of FMIS to understand how precision agriculture can improve the allocation of costs to final products. Finally, we conclude our discussion on the process of adoption of FMIS in European farms.

Keywords Farm management information systems • Precision agriculture • Stakeholders • Adoption

---

Z. Tsiropoulos (✉) • S. Fountas
Agricultural University of Athens, Athens, Greece
e-mail: tsiropoulos@teilar.gr; sfountas@aua.gr

G. Carli
Department of Strategy and Marketing, The Open University Business School, Milton Keynes, UK
e-mail: Giacomo.Carli@open.ac.uk

E. Pignatti
Check Fruit-NSF Italy, Bologna, Italy
e-mail: e.pignatti@nsf-italy.it

S.M. Pedersen, K.M. Lind (eds.), *Precision Agriculture: Technology and Economic Perspectives*, Progress in Precision Agriculture,
https://doi.org/10.1007/978-3-319-68715-5_9

9.1 Introduction to Farm Management Information Systems

The tremendous progress on technological advances in computers and electronics in agriculture in the last decades has brought significant changes in working environment for the farming community. This has generated a vast amount of data to be used by farmers and the challenge is the best exploitation of these data to make useful and practical information available for crop production. The farm manager of today has to choose among different vendors of technologies and data providers to use the most appropriate information to make the best decisions for his or her farm. Decision making is a crucial component for the farmers and many researchers have studied it in relation to the availability of providing data (i.e. Fountas et al. 2006; Magne et al. 2010). The most important aspect of carrying out research in farm management decisions is to understand the tacit knowledge of farmers, and how farmers react when a decision should be made (Gladwin 1989). This is the most important direction that researchers working with data management in agriculture should pursue to provide farmers with the information they need to enhance decision making at specific stages of their production process.

The basis for efficient decision making is availability of high-quality data. In Europe, most of the farms are having difficulties in using the available data and information sources, which are fragmented, dispersed, difficult and time-consuming to use. This indicates that the full potential of these data and information are not well utilized by farmers. The integration of historical data, real-time data from various farming sources, knowledge sources, compliance to standards, environmental guidelines and economic models into a coherent management information system is expected to remedy this situation (Fountas et al. 2005).

Farm management information systems (FMIS) have advanced from simple farm record-keeping systems to large and complex systems in response to the need for communication and data transfer between databases to meet the requirements of different stakeholders. The FMIS are electronic tools for data collection and processing to provide information of potential value in making management decisions (Boehlje and Eidman 1984). They exist when main decision makers use information provided by a farm record system to support their business decision making (Lewis 1998). In a more detailed expression, FMIS is defined as a planned system for collecting, processing, storing and disseminating data in the form needed to carry out farm operations and functions (Sørensen et al. 2010). Essential FMIS components include specific farmer-oriented designs, dedicated user interfaces, automated data processing functions, expert knowledge and user preferences, standardized data communication and scalability; all provided at affordable prices to farmers (Murakami et al. 2007). The FMIS have evolved in sophistication through the integration of new technologies, such as web-based applications and applications for smart phones and tables (Nikkilä et al. 2010).

A key question has been whether commercial FMIS have been able to capture the functionalities developed in academic research, such as an indication of the level of transferal and uptake between research and commercial systems. Another question is whether the increased demands from data intensive Precision Agriculture services is being met by current commercial FMIS systems. Such a comparison between academic with 141 commercial FMIS applications was carried out by Fountas et al. (2015a). Their study revealed that commercial applications mostly deal with data processing for everyday farming activities, whereas academics still explore new horizons in research with high sophistication and complexity, capturing new trends involving spatial and temporal management, distributed systems involving interoperability of sensing devices, future internet components and web services. Commercial applications tend to focus on solving daily farm tasks with the aim to generate income for the farmers through better resource management and field operations planning. The advances that are needed in the development of FMIS include improvements in technology, adaptation motivation, specific new functionalities and greater emphasis on software design governed by usability and human–computer interaction. The diffusion of information management as business innovation in the farming community could benefit from the comprehensive research developed in the last few decades on the adoption of Information and Communication Technologies (ICT) and e-commerce among both consumers and small businesses.

9.2 Farm Management Information Systems Functionalities and Applications

Agriculture is a complex system that incorporates a number of interactions between farmers, advisors, traders, governmental bodies, farm machinery, environmental regulations, economic estimations and others. This system has been summarized in the form of a rich picture in Fig. 9.1 that shows apart from the interactions, the concerns and conflicts between the different entities, where the farm manager is in the middle of the proposed system (Sørensen et al. 2010).

FMIS can cover a large number of functions, such as inventory, calendar, direct sales and site-specific management functions. A set of 10 functions was presented by Fountas et al. (2015a) and is given in Table 9.1.

Apart from human-centered FMIS, there has also been a significant technological evolution in innovations of on-board tractor performance monitoring systems that enables the acquisition of tractor and implement status data through the ISOBUS (universal protocol for electronic communication between implements, tractors and computers) protocol (Tsiropoulos et al. 2013a) and provides useful information to optimize the overall operations and field productivity (Backman et al. 2013). These tractor-based systems together with accurate GPS systems emerge as standard features on contemporary tractors with the aim to provide enhanced farm and operations management through the use of extensive databases as the basis for decision support and control actions. Moreover, the development of autonomous vehicles adopted to field tasks will gradually change the role of the tractor operator toward monitoring and strategic management as this development will require an explicit

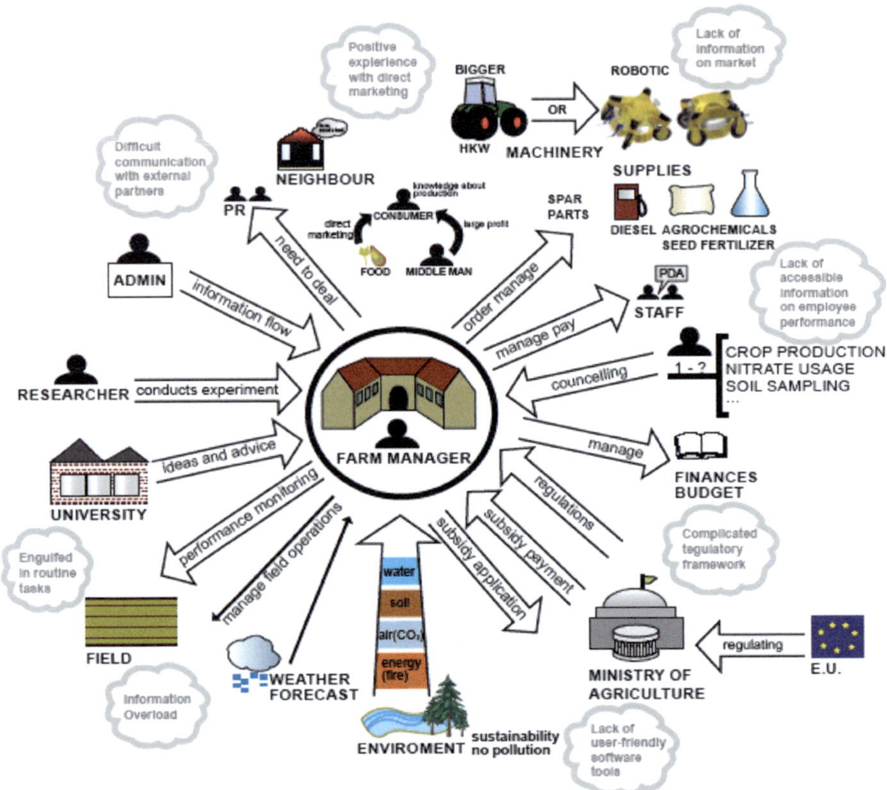

Fig. 9.1 Rich picture of a Farm Management Information System (Sørensen et al. 2010)

management information system capable of managing interactive information flows and provide useful guidelines in real-time for operations execution (Tsiropoulos et al. 2013b). The interconnection between the ISOBUS and precision agriculture innovations will meet the farm manager's demands by open up a wealth of information for improved management of crop production.

With respect to having the tractor in the middle of an information system, a shift of perspective from the farmer or farm manager as the core of the system, to a tractor-centric approach leading to an innovative FMIS architecture where the information flows derive from an intelligent machinery entity that has an upgraded role as part of the decision making process was presented by Fountas et al. (2015b). The term Farm Machinery Management Information System (FMMIS) was used to describe the above approach, which relies on information-to-action decision processes for field operations and is depicted as a rich picture in Fig. 9.2.

However, there is not always a smooth path to commercial availability even for systems that have already shown their potential in a research setting. In just one country, the Netherlands for example, several commercial initiatives to develop geo-information system (GIS) platforms for use in agriculture have failed. However,

Table 9.1 Farm Management Information Systems (Fountas et al. 2015a)

Function title	Function description
Field operations management	Recording of farm activities to help farmer optimize crop production by planning activities and observing the actual execution of planned tasks. Preventive measures may be initiated based on the monitored data.
Best practice (including yield estimation)	Production tasks and methods related to applying best practices according to agricultural standards (e.g. organic standards, integrated crop management (ICM)). A yield estimate is feasible through the comparison of actual demands and alternative possibilities, given hypothetical scenarios of best practices.
Finance	Estimation of the cost of every farm activity, input–outputs calculations, equipment charge-outs, labour requirements per unit area. Projected and actual costs are also compared and input into the final evaluation of the farm's economic viability.
Inventory	Monitoring and management of all production materials, equipment, chemicals, fertilizers, and seeding and planting materials. The quantities are adjusted according to the farmer's plans and customer orders.
Traceability	Crop recall, using an ID labelling system to control the produce of each production section, including use of inputs, employees and equipment, which can be easily archived for rapid recall.
Reporting	Creation of farming reports, such as planning and management, work progress, work sheets and instructions, orders purchases, cost reporting and plant information.
Site-specific	Mapping the features of the field, analysis of the collected data, generation of variable-rate inputs to optimize input and increase output. This is the Precision Farming Technologies component. It could be separate software or integrated.
Sales	Management of orders, charges for services and online sales.
Machinery management	Includes the details of equipment usage, the average cost per work-hour or per unit area. It also includes fleet management and logistics.
Human resource management	Employee management, availability of employees in time and space, handling work times, payment, qualifications, training, performance and expertise.

a system called "Akkerweb" (in English: Farm Maps; www.akkerweb.nl) is currently gaining credence. Akkerweb is the product of a public–private partnership between Agrifirm, the largest farmers' cooperative in The Netherlands, and Wageningen UR, the leading agricultural research organization in The Netherlands.

Akkerweb is geo-information system platform that allows geo-data acquisition, management, visualization and use at the farm level in combination with a standard FMIS (Kempenaar et al. 2016). In addition, farm advisors can access the data if the farmer wants to share data. Akkerweb offers GIS functionality and a number of general free for use applications ("apps"), such as a cropping scheme app, a satellite data app and a sensor data app to visualize and analyze soil and crop data and to generate task maps. Akkerweb also contains several subscription-based apps for variable-rate application of pesticides and fertilizers. The success of Akkerweb is due to the combination of its ICT infrastructure and its science-based content, the bottom-up development with users in the driver's seat, and the effective cooperation

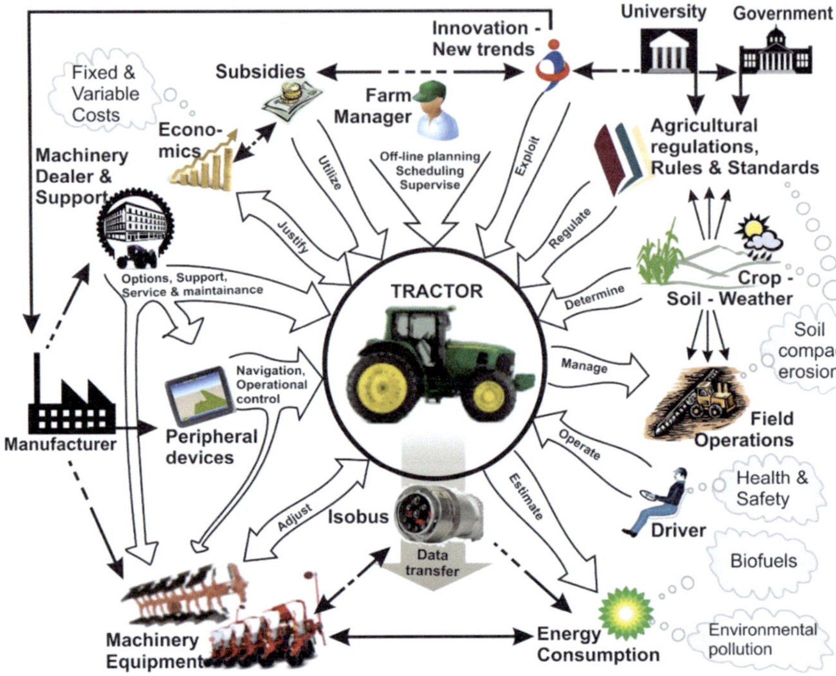

Fig. 9.2 Farm Machinery management information system (Fountas et al. 2015b)

between a farmers' cooperative, a research institute and an IT company with sufficient means to build the required infrastructure. Akkerweb is an open platform in the sense that third parties can also use the Akkerweb platform to develop and offer fee-based services. Today, data of ca. 30,000 crops are stored using Akkerweb.

There are of course many other commercial FMIS in Europe and around the world that are used by farmers or farmers' cooperatives. A successful system is the FARMSTAR in France (https://www.farmstar-conseil.fr), which is a satellite technology-based service devised and delivered by Airbus Defence and Space since 2003. FARMSTAR's users are taking advice on precision agro-management knowing the exact time and area where they should apply fertilizer and pesticides. Satellites flying over the fields take accurate measurements of the radiant solar energy absorbed and reflected from the surface across the farm terrain. The value of the reflected energy varies according to the level of growth of the vegetation, thus satellite measurements can indicate crucial field factors such as soil moisture, surface temperature, leaf cover and level of chlorophyll. Personalized "recommendation cards" divided into very small areas of the field are provided to each user, offering her or him prescriptions for the necessary amounts of chemicals that should be applied, as well as where and when to be applied. The FARMSTAR service provides its subscribers with the opportunity for a better environmental, economic and social management.

9.3 Costing Functionalities of FMIS

One of the main advantages of precision agriculture technologies is to make cost savings in crop production, related to the use of more effective techniques or to reduce the quantity of resources (e.g. water, fertiliser, crop protection) for a range of activities. This advantage has been acknowledged at the level of a single technique to highlight the positive effects of its introduction, but the benefits on the whole farm have received less attention. How a farm might benefit from the use of precision agriculture techniques still remains an open question, given the high initial investment and the level of education and training required. The introduction of evolved FMIS could be seen as a possible answer because they can collect and archive data on the use of resources and elaborate information on final product costing. Moreover, they can provide a more comprehensive picture of the cost of using precision agriculture technologies, evaluating other aspects of the costs of precision agriculture technologies such as the effect of the investment on the final cost of agri-food products.

To support a solid costing functionality, FMIS need a quite sophisticated cost management structure based on three processes (Carli and Canavari 2013): data collection, elaboration of information and decision making. The data collection process is related to these elements depicted in Fig. 9.3: (1) the time spent by human resources on crops, (2) the time spent by machines (e.g. tractor) or equipment (e.g. a precision agriculture device) on each crop, (3) the use of external services in terms of costs and time and (4) the quantity of resource distributed on each crop, in a specific time and position.

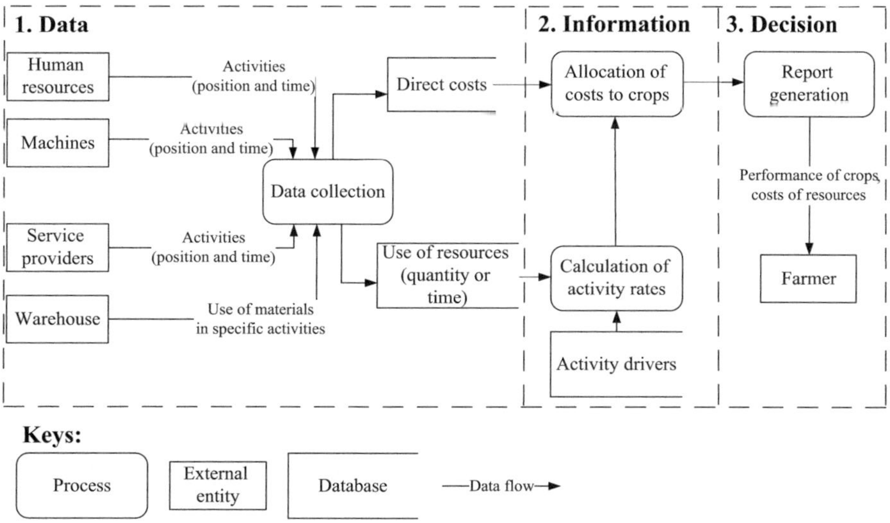

Fig. 9.3 Data Flow Diagram of the cost allocation on crops

In the data collection process, different levels of accuracy can be reached, according to the technological support available. For instance, the use of fertilizer on a crop can be measured as a single value for the whole field without any kind of instrument, or it can be measured more accurately using precision agriculture technology and then drawing a map of its distribution on the field. In this case, the data structure of the FMIS must be designed to track this information as we will discuss later.

The elaboration of cost information aims to provide decision makers such as farmers, technicians, agronomists, with the necessary information on profitability of crops. Decisions on crop production should consider their profitability, but, as anticipated, it reveals that it is particularly complex to collect and elaborate data on costs. Conversely, data on revenues are more accessible because they are based on the market prices of agri-food products or are defined by contractual agreements.

The elaboration phase of cost data can be based on two different models which can be combined together: direct costing and activity-based costing. As an accounting practice, direct costing charges variable costs directly to products (Siegel and Shim 2000). In the case of agricultural practice, this is possible if we charge the direct costs to the activity performed on the specific crop and field. For instance, in a fertilizing activity, the cost of crop protection should be allocated to an activity related to a particular crop (e.g. the second distribution of crop protection on potatoes on field number 2). Although it is quite simple to model an information system to record this type of information, it is far more complicated to record this information from the field, especially when the same activity is carried out on different fields in sequence: for instance, the specific quantity spread on each crop should be recorded. In this case, precision agriculture technologies can provide two types of useful information: (1) the position of the machine or human resource and (2) the quantity of time spent or of resource used. Combining these data, it is possible to adopt a direct costing approach.

If these data are not available, the use of activity-based costing procedures becomes a possible alternative. Activity-based costing methodology has been developed because of the increase in fixed cost share among the total costs of an industrial company (Cooper and Kaplan 1988; Johnson and Kaplan 1987). Its core principle is to allocate fixed costs according to a precise measurement of resource use. First, through the Resource-Activity Assignment Process, the resource consumption generated by the different activities performed in a company is measured; then the Activity-Cost Object Tracing Process finds out which activities are required by products (or final cost objects) and allocates the corresponding portion of costs (Ferreira 2004). The purpose of this paragraph is not to introduce activity-based costing, however, we present a simple example to clarify its logic. Typically, the fixed costs of a tractor could be allocated to crops according to their use. Nevertheless, in the case of a farm with a crop cultivated on a large extension with limited demand for activities involving the tractor, and a crop cultivated on a smaller extension requiring an intense use of the tractor, a classical cost allocation model based on the extension could to be misleading. The large crop would receive the majority of the fixed cost, although it generated a minimal use of the resource, whereas the crop cultivated in the smaller extension would appear to require less resource that it actually did. Although formally correct from an accounting point of view, this procedure could induce an incorrect interpretation on the profitability of the two crops by allocating the majority of the costs to the product with the smaller demand on the activity generating the costs. Conversely, an activity-based procedure for cost allocation could make use of the time spent by the tractor on the two crops. This allocation of cost is able to measure the use of the resource better without producing a significant change in the reality, favouring a consistent process of decision making. To be applied, it is necessary to record the time spent by the tractor on each crop, and then divide its indirect costs (e.g. maintenance, depreciation) according to that time.

The application of precision agriculture technologies could favour the accuracy of the measurement of cost drivers that can be used for activity-based cost allocation. For instance, positioning and mapping solutions could be employed to track human resources, machines and equipment in their movements in the fields. Variable-rate of application systems can record the quantities of material distributed across the field. These two sets of data (position and time spent, and position and quantity of material distributed) could be used as an activity driver to allocate other fixed costs such as depreciation.

Table 9.2 reports a possible solution for cost allocation on final products. In some cases, both the procedures, direct and activity-based are possible and the availability of data determines which is feasible. The time used by machines and human resources can be regarded as the most accessible cost factor as suggested by Kaplan and Anderson (2007).

From this example, the pivotal role of FMIS emerges in supporting the elucidation? of cost data supporting direct costing and activity based costing procedures, and incorporating a reporting functionality dedicated to product costing.

The structure of the FMIS database could be modelled around the entities of fields, crops and activities (Carli and Canavari 2013; Carli et al. 2014). The combination of these elements favour the definition of simple and solid cost allocation procedures. Nevertheless, the advent of precision agriculture technologies can require deep changes in this model: the level of detail reachable with positioning technologies goes far beyond the single field and crop. It is now possible to verify when a machine or a human resource is employed on a specific sub-area of a field or even on a particular tree. This technological evolution enables an even more accurate costing model: for instance, in orchards, the single trees can be considered cost objects, and can be compared in terms of costs and yields (Tsiropoulos and Fountas 2015). This open a new perspective on the modelling of FMIS and the definition of their costing functionality.

Table 9.2 Solutions for cost allocation on final products

Type of resource	Example	Cost	Possibility to apply direct costing and data required	Activity-based costing	Examples of measurement systems
Machine	Tractor	Fuel consumption	Fuel used on a single crop (e.g. level control)	Time spent on each crop	GPS
		Fixed costs incl.: Depreciation	Not applicable		
		Maintenance	Not applicable		
Human resources	Farmer or seasonal worker	Cost per hour	Time spent on a single crop (GPS positioning)	Time spent by human controlled machines on each crop	GPS on machine
Material	Crop protection	Cost of the input	Quantity distributed on each crop (position and quantity from GPS and ISO-BUS)	Time spent by machines on each crop	GPS on machine
	Fertiliser				
	Lime				
	Seeds				
	Water				
External service	Specific crop service (e.g. pruning)	Cost of the service per field	Time spent by external supplier on each crop	Not applicable	Not applicable
	General service (e.g. consulting)	Cost of the service per field	Not applicable	Time spent by human resources or machine son the crop	GPS data

9.4 Adoption of FMIS

The adoption process of technological innovations in agriculture is highly complex because it is affected by a broad range of factors and drivers that could affect the decision to adopt or reject the innovation. Behavioural attitudes, education and awareness, cultural background and norms, social influences, economic and financial variables, policy and market conditions can act as explanatory variables for the adoption patterns of innovation, together with structural and infrastructure factors, availability of support, the characteristics of the innovation itself (Daberkow and McBride 2003; Howley et al. 2012). Examples from literature have proved that the interaction between potential adopters and technologies to be evaluated for adoption must be considered strongly context-specific.

Literature provides examples of models to analyze the set of factors affecting the decision to adopt or reject technological innovations. The Technology Acceptance Model (TAM) (Davis 1989) is widely used in the analysis of the determinants of technology adoption. Focusing on attitude and perception aspects, the model identifies two main constructs (Perceived Usefulness and Perceived Ease of Use) as predictors of the final intention to adopt a technological innovation (User Acceptance). The TAM has been developed further and integrated with constructs from other theoretical models (Venkatesh et al. 2003; Awa et al. 2012). Subsequent adaptations of TAM aimed at identifying the most relevant factors to detect the intention to adopt ICT innovations, both in IT and in the agricultural field (Davis and Venkatesh 2004; Adrian et al. 2005) and tried to validate additional constructs and items to be considered as drivers of the decision process of new technology adoption.

It must be noted, that the strength of factors and drivers affecting farmers' behaviour and their decision to adopt or reject technological innovations depend strongly on many aspects: socio-demographic features of farmers, cultural and social background, characteristics of farms, farming types, type and features of the technology evaluated (e.g. compatibility, costs, profitability, resources savings); external environment (e.g. infrastructure, support from third parties, availability of advisory services, experiences from early adopters, governmental approach, market, financial situation) (Alvarez and Nuthall 2006; Lu et al. 2014; Pierpaoli et al. 2013; Pedersen et al. 2004; Lawson et al. 2011). The relationship between farmers and technologies (e.g. time spent in getting used to the technologies, farmers' dependence on specific solutions and farmers' involvement in the development of new applications) could play a relevant role also in the adoption or rejection choice of technological innovations (Pedersen et al. 2004; Lawson et al. 2011). Finally, requests from stakeholders and actors in the agricultural supply chain (such as traceability or demonstration of environmental sustainability) can exert an influence on farmers' behaviour and decisions (Pedersen et al. 2004).

The use of FMIS in agriculture has been investigated in depth during the last few years because the adoption of management systems to collect and analyze data from in-field activities has become strategically mandatory to support decision-making processes and gain efficiency. The advent of precision agriculture and related technologies provided farmers with large amounts of available data to be processed (Zhang et al. 2002); therefore, information flows and their management, and the consequent support to decision-making are the very critical issues that FMIS must cope with (Sørensen et al. 2010; Fountas et al. 2015a, b).

Many examples of FMIS models can be found in the literature, as outlined in Fountas et al. (2015a, b). During recent years, the development of FMIS has led to the incorporation of more sophisticated functionalities, with the aim of increasing FMIS compatibility with existing technologies, their capability of collecting and processing data, their effectiveness in supporting decision-making. Nevertheless, contributions in the literature have highlighted that their adoption is affected or can be conditioned by some critical factors. Nikkilä et al. (2010) pointed out that usability, reliability, availability, resources saving, convenience, ease of use and connectivity are critical features for end-users when evaluating FMIS. On the other hand, unintuitive or excessively complicated systems, or extremely wide sets of features provided by FMIS could cause misuse and be responsible for low levels of adoption (Nikkilä et al. 2010). Murakami et al. (2007) provided a list of requirements that information systems should possess to support precision agriculture technologies such as integration with existing systems, interoperability with other software packages and data sources, scalability and accessibility. In Sørensen et al. (2010), the interoperability and the transfer of information between systems are mentioned as significant issues to be improved in future FMIS, with the aim of meeting farmers' needs in terms of FMIS functionalities and interfaces. The difficulty in assessing the intangible benefits of information system improvements, and the influence of farmers' computer readiness on the perception about the value of information systems must be included among the critical factors affecting the adoption of FMIS (Alvarez and Nuthall 2006). In addition, other factors such as socio-demographic features of farmers, software fitting and matching with existing systems, ease of use, time and money saving can influence potential users' decisions to adopt FMIS (Alvarez and Nuthall 2006). Similarly, compatibility between hardware and software, adaptability, flexibility, reduction of training needs, and provision of useful and ready-to-use information outputs must be included among the features that FMIS should have to enhance their diffusion (Fountas et al. 2015a, b). Although returns from FMIS adoption in terms of better data management and support to decision making could not be easily quantified by end-users, benefits of the introduction of FMIS should be clearly identifiable and measurable in terms of key performance indicators (Fountas et al. 2015a, b).

Evidence from the literature confirms that advancements and improvements in FMIS design and modelling cannot overlook the interaction with farm stakeholders (Nikkilä et al. 2010), the identification of the scope of a system, boundaries, processes and actors asking for specific requirements of the systems (Sørensen et al. 2010). In the light of these premises, it follows that exploration of the most pertinent factors that affect the intention to adopt FMIS must be deepened, together with a careful evaluation of context-specific variables that could affect farmers' behaviour and perceptions.

Methodologies to estimate FMIS adoption: preliminary exploration of attitudes and beliefs – evidence from the ROBOFARM Project

A study focusing on the identification of the most relevant factors affecting the decision to adopt ICT innovations, and on the steps of this decision process was carried out during the ROBOFARM Project (ICT-AGRI ERA-NET Project "Integrated robotic and software platform as support system for farm level business decisions", funded under the European Union Seventh Framework Programme for Research, Technological Development and Demonstration Activities). This 2-year project aimed to create a demonstrator platform that integrates existing software and hardware technologies into a single system making use of robots with sensors and communication systems to collect data from the field, to be conveyed to and managed by a Farm Management Information System (FMIS).

During the project, a preliminary qualitative analysis was done to understand the attitude of farmers towards ICT innovations and evaluate the adoption of new software solutions for farm information management, together with the relevant steps of the decision process and the intervening factors. Qualitative explorative approaches are usually suggested to conduct in-depth investigations on relatively under-explored topics, trying to identify underlying or latent interactions between factors. Targeting small groups of participants, these methods rely on interviews and focus groups to help in pinpointing the most relevant features of a phenomenon, allowing the identification of significant issues that derive from interviewees' experiences. In the ROBOFARM Project, the focus group discussion method was selected because the fundamental assumption underlying this approach is that opinions, preferences and behaviour emerge from the interaction among informants into a shared context re-created through the focus group setting. Even though focus groups showed some limitations, they enable large amounts of qualitative evidence to be collected, and favour the emergence of experiences and themes (Hines 2000). In particular, they control the interactions and synergy among participants to deepen the investigation of complex behaviour and motivation because the discussion between interviewees provides valuable insight about the extent of consensus and divergence among the group (Morgan 1996).

Six focus groups were established during the summer of 2013 in three countries involved in the project, Greece, Italy and Turkey. A maximum of 10 participants per focus group (recruited among farmers and technicians) were invited to discuss selected topics according to a specific semi-structured protocol aimed at stimulating their interaction. Main topics and objectives of the sections of the qualitative schedule are shown in Table 9.3.

The main objectives of the focus groups were:

- To identify the main factors affecting the decision to adopt a technological innovation (new FMIS);
- To list the steps leading to the adoption of a technological innovation;
- To identify the links between the steps of the process of adoption and the factors that could influence each single step.

Table 9.3 Qualitative schedule of the focus group

Topics (sections)	Objectives
A. Organizational and professional tenure	Role of socio-demographic features (income, company size, years on business, land and equipment ownership, role of the interviewee, age, education) in influencing the adoption of technological innovations
B. Technology adoption in agriculture	Attitudes, opinions and experiences regarding the adoption of technological innovations
C. ICT/technological innovations' adoption process	Identification of the steps that lead to the adoption (or rejection) of technological innovations; identification of the factors (intrinsic and extrinsic) affecting each step of the adoption process
D. Opportunities and limitations	Identification of positive and negative aspects regarding the adoption of technological innovations (benefits, drivers to be enhanced or adjusted, what's missing)

Source: authors' elaboration from (Pignatti et al. 2015)

Results are shown in Table 9.4, which provides a summarized overview on the outcomes of the focus group discussions. A detailed description of the results and main outcomes can be found in (Pignatti et al. 2015).

Interviewees agreed upon a "six-steps" decision process for the adoption of information management technologies in agriculture:

- 1. Identification of needs
- 2. Evaluation of available solutions
- 3. Analysis of scenarios (comparisons of solutions and investments)
- 4. Risks and Benefits analysis and Return on Investments
- 5. Adoption
- 6. Evaluation after use.

Three main groups of factors influencing the adoption decision process were identified during the focus groups.

A. Features of farms and farmers

According to the interviewees, structural features of the farms (e.g. size, income), socio-demographic traits of farmers (age, education) and farmers' perceptions and orientations toward innovation and entrepreneurship are particularly relevant in the first steps of the decision process regarding the adoption of technological innovations, since they can affect the identification of the needs and the evaluation of the available solutions. Then, in the subsequent stages of the decision process (before adoption), additional farmers' features (such as awareness, knowledge gaps, anxiety, uncertainties, familiarity with innovations) were mentioned as particularly influential, as they seem to become relevant when risks/benefits analyses are performed. In these advanced stages of the decision process, economical characteristics of the farms and their development perspectives (both in terms of business and Return of Investment (ROI)) play an important role, because the introduction of new systems for data collection and information management can require significant organizational changes and investments. Availability and provision of training were also mentioned as important factors affecting the decision about adopting innovations: training is fundamental to fill knowledge and experience gaps. Nonetheless it could absorb considerable financial resources and reduce labor hours. Therefore its role in the decision process becomes fundamental especially in the last steps of the process and after the adoption. In fact, being perceived as an investment, training must be available as soon as the innovation is adopted, to make farmers familiar with the new technologies and avoid misuse, inefficiency and rejection.

B. Features of technological innovations

Focus group discussions highlighted the influence of this group of factors on all the steps of the decision process regarding the adoption of new FMIS. In the first stages of the decision process when available solutions are considered, innovations seem to be evaluated according to their "functional" features (such as usability, ease of use, functions, flexibility, reliability). Usefulness was considered by participants as a fundamental feature for ICT innovations during all the stages of the adoption

Table 9.4 Summary of the outcomes of the focus group discussion

	Factors			
		A. Features of farms and farmers	B. Features of technological innovations	C. Features of external environment
Adoption steps	1. Identification of needs	Age	Complexity of needs (short term vs. long term solutions) and of technologies under evaluation	Future growth perspectives
		Education and culture	Type of technology and profitability	Voluntariness/legislation
		Propensity		External/third parties' influence (consultants, technicians, associations)
		Open-mindedness		
		Entrepreneurial orientation		
		Planning orientation		
		Company's size		
		Production type		
		Income/economic status		
	2. Evaluation of available solutions	Age	Ease of use	Third parties' participation to innovations
			Usefulness	
		Open-mindedness	Reliability	Word of mouth and experience sharing (early adopters)
		Perception of risks	Usability	External/third parties' support
		Company's size	Functionality/identifiable performances	
			Flexibility	
			Path dependence from the adopted innovation	
	3. Analysis of scenarios (comparison of solutions and investments)	Anxiety/fear	Usefulness	External/third parties' support
		Awareness raising	Observability of performances	
		Training	Effectiveness	
		Initial investments	Complexity	
		Company's perspectives	Degree of fit and compatibility	
			Trials and tests on the field	
			Perception of costs/benefits	

(continued)

Table 9.4 (continued)

	Factors		
	A. Features of farms and farmers	B. Features of technological innovations	C. Features of external environment
4. Risks/benefits analysis and return on investments	Age	Usefulness	External/third parties' support
	Education	Effectiveness	Financial support
	Anxiety/fear	Perception of costs/benefits	Policies/ legislation
	Familiarity with innovations	Profitability	
	Income/economic status	Price/performance ratio	
	Production type	Path dependence from the adopted innovation	
	Costs and benefits/ ROI		
5. ADOPTION	–	–	–
6. Evaluation after use	Training	Performance	External/third parties' support
		Trials and tests on the field	
		Compatibility	
		Usability	

Source: authors' elaboration from (Pignatti et al. 2015)

process; path dependence from innovations was also mentioned as critical both in the initial and in the latter stages of the decision process, since it could be a constraining factor. When economical evaluations and comparisons become a relevant part of the decision process, additional factors such as effectiveness of the innovation, complexity, degree of fit and compatibility with existing systems, observability of performances, perceived costs and benefits, profitability, and price/performance ratio are taken into consideration. Return on Investments is a pivotal variable that many interviewees mentioned. Insofar as technological innovations might be viable and useful, their evaluation and adoption depends also on their profitability, on investments needed, and on farmers' exposure to risks.

Finally, the fundamental role of trials, field tests, and successful adoption experiences was acknowledged by all the interviewees: in-field demonstrations and cases of pilot farms seem to be a powerful driver to promote the adoption of a technological innovation, and to favor its diffusion among end-users.

C. Features of the external environment

A strong influence of the external environment on adopting technological innovations was acknowledged by interviewees, affecting all the steps of the decision process. Market environment, agricultural policies and legislation, and funding policies define the context in which farmers elaborate on their decision, and exert an unquestionable influence on all the stages of the adoption process. Stakeholders of different nature can orient the decision of adoption and could even force the adoption of specific technological innovations through legislative obligations, or could boost it through supporting measures and economic stimuli. Alternatively they could discourage it controlling different facilitating conditions, such as "innovation-friendly" policy orientations, public funding, and financial support against market risks.

The technological framework surrounding an innovation plays a relevant role: the provision of up-to-date and easy-to-use solutions, along with new approaches for their dissemination (e.g. shareware, open source tools) could promote a faster diffusion of new ICTs, thanks to the reduction of required financial effort and to the availability of affordable solutions.

Word of mouth, sharing of experiences, and contacts with early adopters were listed by participants as influential factors when deciding on the adoption of new FMIS, especially in the first stages of the decision process. Information by pilot farmers, successful or negative experiences of early adopters, and the chance to evaluate concrete results and performances of the innovations seem to be a more reliable reference system for farmers to trust, and to consider when evaluating adoption.

Informants mentioned external support, as a pivotal factor affecting the decision to adopt: qualified external support from technicians, consultants and associations is sought both when available solutions are evaluated, and when the final risks/benefits analysis is performed, since experts' knowledge and experience can increase farmers' awareness and trust toward innovations. External third parties' support can bridge farmers' knowledge gap regarding potential usefulness and profitability of innovations, and enhance their confidence thought demonstrations and trials. Moreover, the involvement of external trusted third parties (such as governments, research institutes, associations) in the development of technological innovations seems to act as a guarantee of reliability of the innovation itself, and increases the likelihood of adopting.

As a conclusion, the results of the focus group discussions of the ROBOFARM project confirmed the importance of well-known factors as influential drivers in the decision process regarding the adoption of new FMIS. Focusing on a specific innovation (new software), some of the factors mentioned in literature were stressed more than others, and some cues for further discussions were provided. The attempt to define the steps of the decision process regarding the adoption of technological innovation and to identify the most relevant drivers affecting each step can be considered a valid suggestion to set up further studies in this area. New research efforts could specify in more detail the crucial steps of the process towards the final decision, and the pertinent factors with the final aim of defining a model of adoption process valid for agribusinesses and able to fill the gaps faced by farmers in assessing new technologies (e.g. knowledge gaps, communication problems, lack of financial support).

The outcomes of the focus group discussions clearly pinpointed that the dynamics underlying the adoption processes of technological innovations are markedly country-specific, "context"-specific, site-specific, technology-specific and farmer-specific. Given this extreme dependency on the context, we advocate further analyses to measure the relative importance of the relevant factors affecting the adoption of technological innovations, and the relations among them (e.g. moderation, mediation) building a theory of adoption specific for the agricultural practice.

9.5 Discussion and Conclusions

A wide range of technologies and tools have become available for capturing, storage, analysis, wireless transmission, visualization, use and sharing of digital data and information in recent years. Several of these technologies are integrated in platforms that facilitate digital data and information use. In addition, farmers collect the data from their daily activities and field operations either through online sensors or manually and in most of the cases at paper format. The necessity to register all activities, as inputs and outputs for farm activities has been enforced by the Cross Compliance requirements by the European Commission. There are a number of software solutions to register these data at farm office, but the ability to gather precise application data at field level does not exist, especially when it is referred to use application of fertilizers and pesticides using modern tractor and implements. This role is expected to be covered by mobile devices that have started to replace computers and in the near future these mobile devices would be the main computational devices for most of computer users. With each passing season, another wave of mobile devices is released, which will be more powerful than the generation preceding it. Mobile devices of today have the necessary processing power, hardware and capabilities for being able to be used efficiently for automated data gathering in the field.

In a recent study on FMIS functions, Fountas et al. (2015a) reviewed 141 commercial FMIS from Europe, North America, and Australia. After defining 11 functionalities that an FMIS can support (see Table 9.1) and verifying their presence in the sample of commercial systems, a cluster analysis was conducted to identify homogenous groups of systems. The cluster analysis revealed four clusters named according to their main features. One of the clusters presented a higher level of complexity supporting functions weakly represented in the systems of the other three clusters. The reason could be that these high level functions–traceability, best-practice estimate, and quality assurance–require the integration of data from different sources (e.g. field and operations, machines, HR). Therefore, they can be deployed only when the overall system reaches a certain level of completeness and complexity.

Two dimensions were identified as the thresholds towards two possible pathways of development of more sophisticated systems. Inventory management makes possible to develop traceability and quality assurance. Site specific functions support the inclusion of decision making functionalities. Future FMIS should go in the direction of combining site specific and inventory management functions in order to collect enough data to convey a reliable support decision making process and solid traceability and quality assurance functions.

Earlier in this chapter, we introduced the FarmBO system as an example of how data from different sources (e.g. machines, HR) can be collected using precision agriculture technologies and generate insights for decision making on costs based on data directly collected on the field (Carli et al. 2014). The availability of site specific functions can favour the collection of more accurate data on costs and the development of more precise analyses on crop costing and profitability.

We envision a promising way for the development of FMIS in the integration of site specific functions into a sophisticated decision making environment, where farmers and technicians are provided with reports to improve their choices and increase the yields of their crops. This would be possible only if data from sensors are processed with well-established cost management approaches adapted to the specificities of the agricultural practice. For instance, since site-specific solutions applied to orchards may offer data on the single trees very soon, the amount of crop protection would be decided and measured for each single tree. The integration of precision agriculture solutions and the decision support module of a FMIS can pave the way to a more fine grained accounting process till the level of the single tree. New research efforts could be dedicated to the definition of a straightforward stepwise process to elaborate the rich and complex data from sensors. Therefore, the decision support module of the FMIS would be able to provide farmers with just the relevant data for each activity and choice to make. A challenge for future FMIS is in this meso-level of data elaboration: only the systems able to *make sense* of the richness of the data provided by sensors and *advise* the farmer on possible options will differentiate in the competitive arena.

References

Adrian AM, Norwood SH, Mask PL (2005) Producers' perceptions and attitudes toward precision agriculture technologies. Comput Electron Agric 48(3):256–271

Alvarez J, Nuthall P (2006) Adoption of computer based information systems: The case of dairy farmers in Canterbury, NZ, and Florida, Uruguay. Comput Electron Agric 50(1):48–60

Awa HO, Ojiabo OU, Emecheta BC (2012) Integrating TAM and TOE frameworks and expanding their characteristic constructs for E-commerce adoption by SMEs. In: Proceedings of Informing Science & IT Education conference (InSITE) 2012 (12:571–588). Informing Science Institute

Backman J, Oksanen T, Visala A (2013) Applicability of the ISO 11783 network in a distributed combined guidance system for agricultural machines Original Research Article. Biosyst Eng 114(3):306–317

Boehlje MD, Eidman VR (1984) Farm management. Wiley, New York, p 806

Carli G, Canavari M (2013) Introducing direct costing and activity based costing in a farm management system: a conceptual model. Procedia Technol 8:397–405

Carli G, Canavari M, Grandi A (2014) Introducing activity-based costing in farm management. Int J Agric Environ Inf Syst 5(4):69–84

Cooper R, Kaplan RS (1988) Measure costs right: make the right decisions. Harv Bus Rev 66(5):96–103

Daberkow SG, McBride WD (2003) Farm and operator characteristics affecting the awareness and Adoption of precision agriculture technologies in the US. Precis Agric 4(2):163–177

Davis FD (1989) Perceived usefulness, perceived ease of use, and user acceptance of information technology. MIS Q 13(3):319–340

Davis FD, Venkatesh V (2004) Toward preprototype user acceptance testing of new information systems: implications for software project management. IEEE Trans Eng Manag 51(1):31–46

Ferreira WN (2004) Is activity based costing a blockbuster? extension economic reports. Clemson University, Clemson

Fountas S, Ess D, Sorensen CG, Hawkins S, Blumhoff G, Blackmore S, Lowenberg-DeBoer J (2005) Farmer experience with precision agriculture in Denmark and the US eastern corn belt. Precis Agric 6:121–141

Fountas S, Wulfsohn D, Blackmore S, Jacobsen HL, Pedersen SM (2006) A model of decision making and information flows for information-intensive agriculture. Agric Syst 87:192–210

Fountas S, Carli C, Sørensen CG, Tsiropoulos Z, Cavalaris C, Vatsanidou A, Liakos B, Canavari M, Wiebensohn J, Tisserye B (2015a) Farm management information systems: current situation and future perspectives. Comput Electron Agric 115:40–50
Fountas S, Sorensen CG, Tsiropoulos Z, Cavalaris C, Liakos V, Gemtos T (2015b) Farm machinery management information system. Comput Electron Agric 110:131–138
Gladwin H (1989) Ethnographic decision tree modelling. Sage Publications Ltd., London
Hines T (2000) An evaluation of two qualitative methods (focus group interviews and cognitive maps) for conducting research into entrepreneurial decision making. Qual Mark Res Int J 3(1):7–16
Howley P, Donoghue CO, Heanue K (2012) Factors affecting farmers' adoption of agricultural innovations: a panel data analysis of the use of artificial insemination among dairy farmers in Ireland. J Agric Sci 4(6):171
Johnson HT, Kaplan RS (1987) Relevance lost: the rise and fall of management accounting. Harvard Business Press, Boston
Kaplan RS, Anderson SR (2007) Time-driven activity-based costing: a simpler and more powerful path to higher profits. Harvard Business Press, Boston
Kempenaar C, van Evert FK, Been T, Kocks CG, Westerduk CE (2016) Towards data-intensive, more sustainable farming: advances in predicting crop growth and use of variable rate technology in arable crops in the Netherlands. ICPA, St. Louis
Lawson LG, Pedersen SM, Sørensen CG, Pesonen L, Fountas S, Werner A, Oudshoorn FW, Herold L, Chatzinikos T, Kirketerp IM, Blackmore S (2011) A four nation survey of farm information management and advanced farming systems: a descriptive analysis of survey responses. Comput Electron Agric 77:7–20
Lewis T (1998) Evolution of farm management information systems. Comput Electron Agric 19:233–248
Lu Y, Lu Y, Wang B, Pan Z, Qin H (2014) Acceptance of government-sponsored agricultural information systems in China: the role of government social power. IseB 13(2):329–354
Magne MA, Cerf M, Ingrand S (2010) A conceptual model of farmers' informational activity: a tool for improved support of livestock farming management. Animal 4:842–852
Morgan DL (1996) Focus groups. Annu Rev Sociol 22:129–152
Murakami E, Saraiva AM, Ribeiro Junior LCM, Cugnasca CE, Hirakawa AR, Correa PLP (2007) An infrastructure for the development of distributed service-oriented information systems for precision agriculture. Comput Electron Agric 58(1):37–48
Nikkila R, Seilonen I, Koskinenet K (2010) Software architecture for farm management information systems in precision agriculture. Comput Electron Agric 70(2):328–336
Pedersen SM, Fountas S, Blackmore BS, Gylling M, Pedersen JL (2004) Adoption and perspectives of precision farming in Denmark. Acta Agric Scand Sect B – Soil Plant Sci 54(1):2–8
Pierpaoli E, Carli G, Pignatti E, Canavari M (2013) Drivers of precision agriculture technologies adoption: a literature review. Procedia Technol 8:61–69
Pignatti E, Carli G, Canavari M (2015) What really matters? a qualitative analysis on the adoption of innovations in agriculture. J Agric Inf 6(4):73–84
Sørensen GC, Fountas S, Nash E, Pesonen L, Bochtis D, Pedersen SM, Basso B, Blackmore SB (2010) Conceptual model of a future farm management information system. Comput Electron Agric 72:37–47
Siegel JG, Shim JK (2000) Accounting handbook. Barron's Educational Series, Hauppauge
Tsiropoulos Z, Fountas S (2015) Farm management information system for fruit orchards. Precis Agric 15:217–232
Tsiropoulos Z, Fountas S, Gemtos T, Gravalos I, Paraforos D (2013a) Management information system for spatial analysis of tractor-implement draft forces. Precis Agric 13:349–356
Tsiropoulos Z, Fountas S, Liakos V, Tekin B, Aygun T, Blackmore S (2013b) Web-based farm management information system for agricultural robots. EFITA-WCCA-CIGR Turin, Italy
Venkatesh V, Morris MG, Davis GB, Davis FD (2003) User acceptance of information technology: toward a unified view. MIS Q 27(3):425–478
Zhang N, Wang M, Wang N (2002) Precision agriculture—a worldwide overview. Comput Electron Agric 36(2–3):113–132

Chapter 10
Sustainable Intensification in Crop Farming – A Case from Estonia

Rando Värnik, Raiko Aste, and Jelena Ariva

Abstract Sustainable intensification is a theoretical approach under development that takes into account the economic, social and environmental aspects of production. So far little attention has been given to assess the variables that have an impact on sustainable intensification in the field of agriculture in Estonia. The current study is based on data from 119 agricultural companies in 2012. The application of different technologies used in crop production was analysed and many previously presented factors from literature about sustainable intensification were included in this research. Findings from this study show that both the selection of inputs for production as well as age and education have an impact on yields of spring wheat. This refers to the fact that the yield made by younger farmers is higher primarily because of updated knowledge and willingness to test out appropriate inputs and technologies selected for production. The research also shows that half of the producers that were analysed apply sustainable intensification in agriculture to some extent.

Keywords Sustainable intensification • Regression analysis • Wheat production technology • Environment

10.1 Introduction

An increase in the world population and food demand in addition to a greater attention to agriculture-related environmental problems (FAO 2014) have presented a variety of challenges for the food production sector, including questions such as how to produce more from the limited resources while doing it efficiently and in an environmentally friendly way (TRS 2009; SDSN 2013; Cook et al. 2015). In addition to the concept of sustainable agriculture, the concept of sustainable

R. Värnik (✉) • R. Aste • J. Ariva
Estonian University of Life Sciences, Institute of Economics and Social Sciences, Tartu, Estonia
e-mail: rando.varnik@emu.ee; raiko.aste@emu.ee; jelena.ariva@emu.ee

S.M. Pedersen, K.M. Lind (eds.), *Precision Agriculture: Technology and Economic Perspectives*, Progress in Precision Agriculture,
https://doi.org/10.1007/978-3-319-68715-5_10

intensification, which takes into account the economic, social and environmental aspects, has recently come to the fore as a response to ensuring food security for the growing world population (Buckwell et al. 2014). The two concepts 'sustainable agriculture' and 'sustainable intensification 'are used interchangeably in literature, and the distinction between these two terms may be blurred. The term 'sustainable intensification' concentrates on raising productivity while reducing the negative environmental impacts and increasing environmentally-friendly land use. Sustainable intensification not only focuses on a single activity, but also studies the broader patterns behind the changes in productivity per unit of area and the environmental impact of land use, together with the changes from the initial state or situation (Barnes and Thomson 2014).

The term 'sustainable intensification' originates from the 1990s when it was introduced in relation to agricultural production in Africa that was characterised by very low yields and ongoing environmental degradation. Sustainable intensification was defined as an attempt to produce in such a way that increases in yields would not give rise to adverse environmental impacts and cultivation of more land (Garnett and Godfray 2012). Originally this concept was related to agricultural production in developing countries, but it has now spread all over the world and the term is subject to a range of different interpretations.

The US Government initiative "Feed the Future" (USG 2015) defines 'sustainable intensification' as a new approach, which enables the world's growing demand for food to be satisfied. Such an approach builds on the use of modern technologies that have been designed to understand how these technologies work within local agro-ecological systems in order to enhance productivity. Furthermore, to stimulate economic growth and ensure the viability of agriculture by using fewer natural resources as well as ensuring the health and well-being of all the livestock species managed (Garnett and Godfray 2012; SDSN 2013; USG 2015). Dillon et al. (2014) describe sustainable intensification as an increase in desired outputs with the same or fewer inputs, but with significantly reduced or eliminated environmental degradation. Sustainable intensification is not the same for all farming systems, and the necessary actions and developments are in part dependent on the productivity of the current agricultural system and the environmental performance of the farm. This may be reflected in an increase in farm output against the environmental services per hectare or an increase in agricultural output per hectare, which means that in addition to improving productivity, concurrent environmental management is of importance in sustainable intensification.

To this end, knowing how to combine and manage material inputs, i.e. following the principle of "more knowledge per hectare", takes a central role, (Buckwell et al. 2014). Campbell et al. (2014) see a close link between sustainable intensification and climate smart agriculture, whereas the former has the leading role in climate change adaptation (and mitigation). This will potentially result in even lower emissions per unit of production. Climate smart agriculture focuses on the outcomes that are related to climate change adaptation and mitigation, and will always be a part of sustainable agriculture. However, both are only a part of the multifaceted approach that includes the reduction of consumption and waste, creates social guarantees, facilitates trade and improves nutrition (Campbell et al. 2014; Cook et al. 2015). Thus, sustainable intensification in agriculture produces changes throughout the whole food supply chain from production to consumption, including changes in the behaviour of both consumers and food manufacturers (SDSN 2013).

In summary, it can be said that sustainable intensification is not a specific method of production. Sustainable intensification is a continuous process that takes into account the existing resources and productive environment. In general, it aims at producing enough food while maintaining both the agricultural environment and preserving a favourable living environment for future generations. In crop production, for example, this means an increase in yield per inputs (nutrients, water, energy, capital and land), as well as a decline in the negative external effects per unit (greenhouse gas emissions, groundwater pollution) (Garnett and Godfray 2012).

Despite the positive aspects of sustainable intensification, this approach has recently been faced with a lot of criticism. First of all, there are some concerns that sustainable intensification will focus on intensification, as was the case, for example, after the industrial revolution or the Green Revolution, whereas environmental impacts may be relegated. This, however, will lead to environmental damage (impacts on soil, water, air quality and biodiversity), and to a decline in agricultural sustainability because more and more resources (both inputs and non-renewable resources) are used. Some studies, however, indicate that high productivity agriculture may reduce the negative impacts on biodiversity and greenhouse gas emissions (Garnett and Godfray 2012; Ariva et al. 2015; Schiefer et al. 2015).

When focusing only on minimizing the negative environmental impacts, the effects that increased food production have had on improving people's well-being, which is one of the main goals of this approach, may be overlooked. In addition, it is also believed that when concentrating strictly on food production, sustainable intensification is not sufficient for improving food security because a solution to this problem requires a comprehensive approach, which also addresses availability of food (fair distribution of food and personal empowerment). At present, the concept of sustainable intensification does not include the principle of fairness, but in many cases, food security problems can be solved through the improvement of justice and fairness, (Loos et al. 2014; Cook et al. 2015).

Because of the different approaches and definitions, it is not clear which specific agricultural production practices or technologies can be regarded as characteristic for sustainable intensification. Because sustainable intensification is a long-term process with site-specific objectives that often lack clear measurable indicators or criteria, their precise measurement is difficult, especially at the enterprise level (Dillon et al. 2014). Many studies (TRS 2009; Kassam et al. 2011; Barnes 2012; Garnett and Godfray 2012; AFI 2013; Elliott et al. 2013; Buckwell et al. 2014; Dillon et al. 2014; Drechsel et al. 2015; Huggett 2015; Lampkin et al. 2015; Smith et al. 2015) have used indicators that allow the dynamics or the direction of the process from a sustainable intensification theory point of view to be assessed. The analysis of various single impact factors (indicators) of sustainable intensification allows a better identification of their effect on the development of agriculture, but very often their complex or integrated impact is even more essential. Figure 10.1

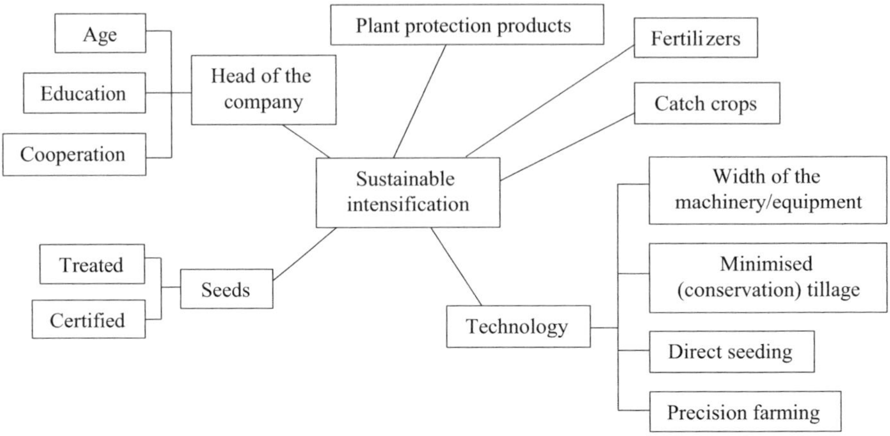

Fig. 10.1 Variables that have an impact on sustainable intensification
Source: Aste (2016), TWB (2005), TRS (2009), Kassam et al. (2011), Barnes and Poole (2012), AFI (2013), Elliott et al. (2013), Buckwell et al. (2014), Dillon et al. (2014), Drechsel et al. (2015), de Haan and Setshwaelo (2015), Huggett (2015), Lampkin et al. (2015), Pretty and Bharucha (2015), and Smith et al. (2015)

highlights the most relevant sustainable intensification indicators as identified in the literature that illustrates the importance of various factors in the system.

The combination of the managers' age and education combined with teamwork skills affects the managers' management decisions that are made regarding inputs (choice of pesticides, fertilizers, seeds) and technologies (cover crops, machinery, tillage practices), sets of values and attitudes, as well as level of innovation.

Based on the above, it is development and cooperation, innovation and technology, genetic diversity (including genetically modified organisms) and agro-ecological intensification and agro-ecological environment that foster sustainable intensification in agriculture (Aste 2016). Collaboration between the public and private sectors in research and development activities helps to boost agricultural productivity, but the cooperation will also enhance environmental awareness among farmers (e.g. allows a better understanding of the links between the environmental needs and land use decisions) and improve their environmental performance (Buckwell et al. 2014). Research and development activities are closely related to technology and innovation, which in limited circumstances (scarce land and water resources) play an important role in increasing productivity. Innovations (such as the use of ICT in agriculture) contribute to the quality of crop production, livestock health and the farmers' quality of life. Machines and equipment become "smarter" and more compatible with the needs, and increase the efficiency of agricultural inputs and outputs. In addition, the advances in technology make the collection of bulk data from various sensors possible, which in turn provides for analyses and subsequent optimized decision-making at every stage of the food supply chain (Hogan 2015).

Conservation of genetic diversity (growing different crop species and varieties, keeping different breeds of farm animals separately) helps to ensure resilience to pests and diseases. Biotechnology (tissue cultures, genomics, molecular selection, genetic engineering) may improve the efficiency of conventional plant and livestock production, allowing for a better understanding and use of natural genetic diversity while reducing the negative environmental impacts (for example, due to reduced use of pesticides). The application of transgenic or genetically modified organisms (GMO) may be useful if the available natural variation of the gene pool is not sufficient to overcome the major obstacles in raising crop and livestock productivity, improve their disease tolerance and the quality of nutrition. Agro-ecological intensification in crop production is mainly related to the application of good agronomic governance principles in local production conditions, which increases profitability (for example, through maximizing yield) and reduces pressure on the environment (Garnett and Godfray 2012; SDSN 2013; Buckwell et al. 2014).

The close and reciprocal interconnection between agricultural activity and the environment leaves a rather negative mark on nature: biodiversity is lost, which upsets the balance in the ecosystem; greenhouse gas emissions affect the climate; tampering with plant nutrient content leads to the pollution of the aquatic environment (surface and groundwater); water consumption in agriculture is putting pressure on the existing water resources and soil degradation reduces soil fertility (Garnett and Godfray 2012; Buckwell et al. 2014). In addition to the adverse environmental impacts (such as higher than average nitrogen surplus per hectare), agricultural intensification may also have some positive impacts. For example, in Ireland more intensive top producing dairy farms emit less greenhouse gases and are economically more successful than their less intensive colleagues (Dillon et al. 2014). The forecast analysis of the dairy industry that was carried out in Estonia also showed that the increase in milk yield at the expense of dairy cow productivity reduced greenhouse gas emission per kg of milk produced. On the other hand, growth in productivity may cause health problems in the herd and shorten the productive lifetime of dairy cows. It is essential that productivity growth is achieved through selection as a result of natural physiological processes (Ariva et al. 2015).

The need to combat the undesirable environmental impacts of EU intensive agriculture and improve resource efficiency and sustainability of agriculture led to an analysis of sustainable intensification on a global scale (the issues that were addressed included plant nutrient recovery and recycling and biodiversity management). The analysis provided an estimate that around 41% of the arable land in the 25 EU member states (excluding Romania, Bulgaria and Croatia) is suitable and 12% could be suitable for sustainable intensification. It emerged that 47% of the arable land was deemed unfit for intensification and extensive agriculture was recommended for 4% of the current arable land. Out of the 3822.8 km^2 of arable land (or 58% of total arable land) analysed in Estonia, 65% is suitable for sustainable intensification (including 0.1%, which is suitable with restrictions), whereas 35% is not suitable (including 0.5%, where extensification is recommended), (Buckwell et al. 2014).

Apart from the analysis prepared by the EU and mentioned above, no Estonia-wide research into sustainable intensification had been undertaken before, and the theoretical basis for research in this area is lacking, which means that there is no overview of the level of sustainable intensification of agriculture in Estonia (Aste

2016). As this issue is particularly important from the environmental point of view and with stricter requirements are imposed on farmers, a study was drawn up that could shed light on the practices and results of sustainable intensification in a specific area of Estonia. Extrapolating from one-year data, the following research aims to give a statistically sound overview of sustainable intensification practices applied in Estonian crop production based on a case study.

10.2 Materials and Method

This study is based on a survey conducted by the Institute of Economics and Social Sciences of the Estonian University of Life Sciences in 2012. The plant production questionnaire consisted of 95 questions, which attempted to map the crop production technologies used, and to identify the selection of inputs and management decisions. The 333 respondents included farms of different sizes, active in either organic or conventional crop production. Most of the companies did not reply to the questionnaire in full, which is why the study is based on the data from only 119 respondents who provided the necessary data for this study.

Based on the data of the survey, the level of sustainable intensification of the crop production farms was assessed according to the indicators outlined in theory. A quantitative research method (correlation and regression analysis) was used to analyse the results. The analysis was performed taking spring wheat as an example because wheat constitutes an important part of Estonia's crop production (26.7% in 2015 in monetary value), and spring wheat acreage has been accounting for more than one-fifth of the cereals acreage in the past 10 years (Statistic Estonia (SE) 2016). The analysis is based on the most important sustainable intensification indicators (variables) that allow for comparison of different technologies. The following indicators were chosen: yields, types of seeds used, application of plant protection products, mineral fertilizers and organic fertilizers. The indicators mentioned above were divided further into sub-categories. A detailed overview of the categories and subcategories of inputs used in wheat production is presented in Table 10.1.

Based on the technology applied, all agricultural enterprises in the survey were divided into three groups. Farms using direct seeding constituted the first group. Data from the literature classifies direct seeding as a sustainable intensification technology (FAO 2011). Farms belonging to the second group practice conservation tillage (use of minimum tillage), which according to literature, can also be regarded as a sustainable intensification technique. The third group consisted of companies that used ploughing as a means of cultivation, which, based on the literature, cannot be considered as a sustainable intensification technology.

To perform the analysis, the necessary source data (indicators) were coded using dummy variables, which took the value of 1 to indicate the presence of the assigned indicator, and 0 if the indicator was not used. Different sustainable

Table 10.1 Wheat production inputs

Seeds used		Use of plant protection products		Use of mineral fertilizers		Use of organic fertilizers	
Own produced seeds	Purchased seeds	Yes	No	Yes	No	Yes	No
Certified	Certified			Liquid fertilizer		Liquid manure	
Uncertified	Uncertified			Granular fertilizer		Solid manure	
Certified dressed	Certified dressed						
Certified undressed	Certified undressed						
Uncertified dressed	Uncertified dressed						
Uncertified undressed	Uncertified undressed						
Dressed	Dressed						
Undressed	Undressed						

Source: Aste (2016)

intensification indicators constituted the independent variables in the regression analyses. Codes x_1–x_{33} were attributed to the variables used, i.e. a total of 33 variables was used in the correlation analysis (Appendix A). The links between the characteristics of farms were studied focusing particularly on the interdependence between the sustainable intensification indicators or variables and yields (Appendix B).

A separate regression analysis was performed for all the different technologies used (direct seeding, conservation tillage, ploughing). As a result, three regression equations were obtained, where spring wheat yields acted as dependent variable (Y) and one of the technology variables and seven additional variables selected from the correlation analysis constituted the independent variables (x).

In addition, the relations between age and educational level of farm managers, the application of precision farming technologies and catch crop cultivation, and the yield and sustainable intensification were analysed.

10.3 Results

The regression analysis showed that in the case of direct seeding of spring wheat, sustainable intensification indicators explain (R^2) 33% of the variation in the yield (Appendix C). The following regression equation was drawn up on the basis of the regression analysis of direct seeding (10.1):

$$Y = 2.2516 - 0.2214x_2 - 0.8543x_9 + 0.3678x_{10} + 0.3605x_{20} \\ -0.8675x_{21} + 1.1462x_{26} + 0.0450x_{24} + 0.8966x_{30} \quad (10.1)$$

where x_2 is direct seeding technology, x_9 purchased uncertified seed, x_{10} purchased certified dressed seed, x_{20} own produced uncertified dressed seed, x_{21} own produced uncertified undressed seed, x_{24} plant protection products were used, x_{26} fertilizers were used, x_{30} organic liquid manure was used.

In the case of direct seeding, the regression equation obtained shows that when using purchased uncertified seeds or own produced uncertified undressed seeds, the yields of spring wheat go down by 221 kg ha^{-1}, 854 kg ha^{-1} and 868 kg ha^{-1}, respectively. The use of purchased certified dressed seeds and own produced uncertified dressed seeds, as well as the use of fertilizers, plant protection products and organic liquid manure increases spring wheat yields by 368 kg ha^{-1}, 361 kg ha^{-1}, 1146 kg ha^{-1}, 45 kg ha^{-1}and 897 kg ha^{-1}, respectively.

In respect of the application of minimized tillage technology, the regression analysis showed that the selected indicators of sustainable intensification explain 34% of the variation in the yield (Appendix C). On the basis of the regression analysis, the following regression equation was generated (10.2):

$$Y = 2.0594 + 0.3243x_3 - 0.7536x_9 + 0.3688x_{10} + 0.3337x_{20} \\ -0.7992x_{21} + 1.3178x_{26} - 0.1177x_{24} + 0.8397x_{30} \quad (10.2)$$

where x_3 conservation tillage.

The regression equation revealed that conservation tillage raised spring wheat yields by 324 kg ha^{-1}. The application of purchased certified dressed seeds (increase of 369 kg ha^{-1}), own produced uncertified dressed seeds (increase of 1318 kg ha^{-1}) and organic liquid manure (increase of 840 kg ha^{-1}) also had a positive effect on the spring wheat yields.

The use of both purchased uncertified seeds and own produced uncertified undressed seeds lowers the yield of spring wheat by 754 kg ha^{-1}and 799 kg ha^{-1}, respectively.

As to ploughing technology, the regression analysis established that the selected sustainable intensification indicators explain 33% of the variation in the spring wheat yield (Appendix C). The following regression equation was drawn up (10.3):

$$Y = 2.3052 - 0.2110x_4 - 0.7785x_9 + 0.3900x_{10} + 0.3567x_{20} \\ -0.8352x_{21} + 1.2553x_{26} - 0.0809x_{24} + 0.8489x_{30} \quad (10.3)$$

where x_4 ploughing.

In relation to the regression equation, ploughing technology reduces the yield of spring wheat by 211 kg ha^{-1}. The use of purchased uncertified seeds, own produced uncertified undressed seeds, as well as plant protection products decreases the yields by 779 kg ha^{-1}, 835 kg ha^{-1}and 81 kg ha^{-1}, respectively. At the same time, the application of purchased certified dressed seeds and own produced uncertified dressed

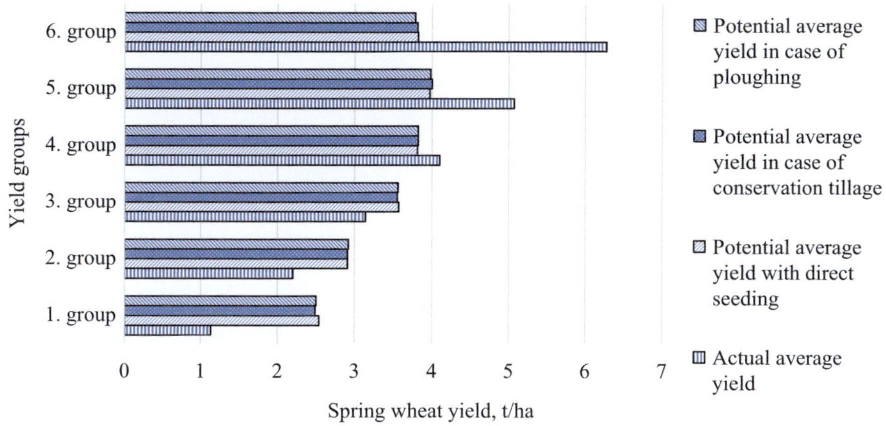

Fig. 10.2 Average actual and potential yields of spring wheat per different technologies used, t ha^{-1} (Source: Aste 2016)

seeds increase the yields by 390 kg ha^{-1}and 357 kg ha^{-1}. Furthermore, the use of fertilizers and organic liquid manure also increase spring wheat yields (by 1255 kg ha^{-1}and 849 kg ha^{-1}, respectively).

Based on the coefficients of the regression equation, potential yields have been calculated for different technologies (Fig. 10.2). Potential yields are divided into six groups. Eight enterprises constitute the first group (yield up to 1.5 t ha^{-1}). Group 2 (yield1.6–2.5 t ha^{-1}) is made up of 21 farms, Group 3 (yield 2.6–3.5 t ha^{-1}) of 32, Group 4 (yield 3.6–4.5 t ha^{-1}) of 38, Group 5 (yield 4.6–5.5 t ha^{-1}) of 15, and Group 6 (yield 5.6–7.3 t ha^{-1}) of 5 companies.

In the first group, the average observed yield of spring wheat is 1.13 t ha^{-1}. In the case of direct seeding and selected farm inputs (indicators) the potential yield is 2.54 t ha^{-1}, which is 124.8% higher than the actual yield. In the case of ploughing, the potential yield is 1.37 t ha^{-1} (121.2%) higher than the actual yield. Although ploughing does not fall into the category of sustainable intensification technology, higher yields are mainly attributable to the use of other sustainable intensification technologies and inputs. In the case of conservation tillage, the potential yields in this group exceed the actual yields by 1.36 t ha^{-1} (120.4%).

The average observed yield in the second group is 2.20 t ha^{-1}. In this group, the use of ploughing has the greatest potential, increasing the yield by 0.72 t ha^{-1} (32.7%) compared to the actual yields. The impacts of conservation tillage and direct seeding technologies on the potential yield are similar to ploughing.

The differences between the actual and potential yields are smallest in the third group. Direct seeding that is regarded as a sustainable intensification technology ensures the highest yields. It exceeds the actual yields by 0.43 t ha^{-1} (13.7%). In the case of ploughing, the potential yields surpass the actual yields by 0.42 t ha^{-1} (13.4%), and in case of conservation tillage by 0.41 t ha^{-1} (13.1%).

The observed yields in group 4 are higher than the potential yields, a result that can be achieved by combining different inputs and technologies. This provides evidence of the situation where the technologies (inputs) applied in the agricultural enterprises are already sufficiently well-combined, and an additional increase in yields can be achieved only by modifying additional inputs or making other choices. The average actual yield is 4.1 t ha^{-1}, whereas the potential yields in case of direct seeding and ploughing would be by 0.28 t ha^{-1} (7.3%) lower, and in case of conservation tillage by 0.29 t ha^{-1} (7.1%) lower than the actual yield.

In group 5, the difference between the actual yield and the potential yields achieved with various technologies was significantly wider. For conservation tillage, the potential yields fall short of the actual yields by 1.07 t ha^{-1} (21.1%) and for ploughing and direct seeding by 1.09 t $ha^{-1}$and 1.10 t ha^{-1}, respectively.

In the sixth group, the difference between the actual yields and the potential yields to be achieved by the application of different technologies was the greatest. If direct seeding is applied, actual yields exceed the calculated potential yields by 2.46 t ha^{-1}. The same figures for conservation tillage and ploughing are 2.47 t $ha^{-1}$and 2.50 t ha^{-1}, respectively.

According to the literature sources mentioned above, precision farming technology is one of the indicators of sustainable intensification. Figure 10.3 shows that in the group where spring wheat yields are up to 1.5 tons per hectare, only 37.5% of the companies are engaged in precision farming. In the next group where the yield is 1.6–2.5 tons per hectare, 47.6% of the companies use precision farming, which is 10.1% more than in the first group. In the third group, half (50%) of the farmers are engaged in precision farming. Although the difference with the previous group is only 2.4 percentage points, it is evident that in the case of higher yields more emphasis is placed on precision farming technologies. In Group 4, precision farming is used by more than a half of the companies, specifically by 63.2%. In Groups 5 and 6, or groups with the highest yields, 80% of the farmers apply precision farming technologies. Thus, it follows that the application of precision farming technologies by farms

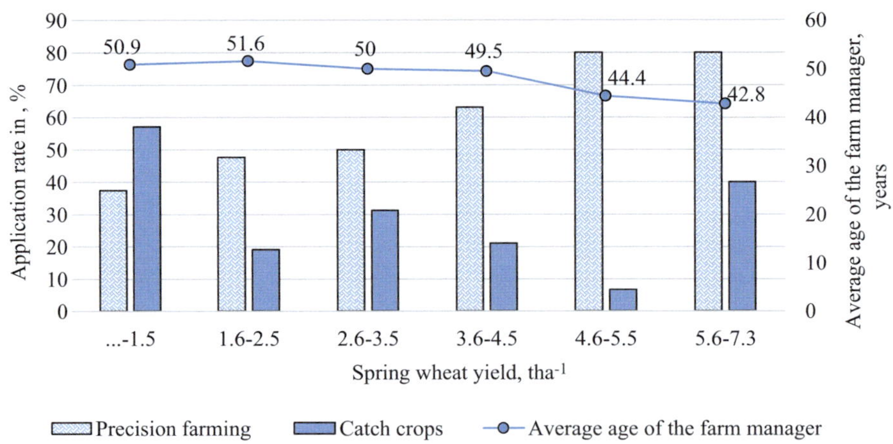

Fig. 10.3 Average age of farm managers and the application of precision farming technology and catch crops by groups (Source: Aste 2016)

in the groups with lower yields could allow the farms to increase their yields per hectare considerably, and in this way move towards sustainable intensification.

The age of the manager is also regarded as one of the indicators of sustainable intensification. Although older entrepreneurs are expected to be smarter and wiser, the yields do not reflect this. Among the groups analysed, the average age of the managers was the highest (approximately 52 years) in the group with yields from 1.6–2.5 t ha^{-1}, and the lowest (about 43 years) in the group with yields from 5.6–7.3 t ha^{-1}. Thus, it appears that younger managers are more willing to adopt new technologies and apply them effectively in production.

The analysis of catch crop cultivation showed that the results obtained from the sample are not consistent with the theory, stating that for sustainable intensification catch crops must be grown to increase species diversity. In group 1 where the yield per hectare is 1.5 t and where farms are not engaged in sustainable intensification, catch crop cultivation is the highest, 57%. In group 2 (yields 1.6–2.5 t ha^{-1}), intercropping was used in 19% of the farms, and in group 3 (2.6–3.5 t ha^{-1}) in 31% of farms, respectively. In groups 4 and 5, the percentage of catch crop cultivation decreases and intercropping is practiced on 21% of the farms, and in the group with yields between 4.6–5.5 t ha^{-1}in 7% of the farms only. The exception is the group with the highest yields, where catch crop cultivation is practiced on 40% of the farms. However, this situation could, to some extent, be explained by the EU support schemes, which encourage intercropping on smaller farms. Larger production units use catch crop cultivation consciously for crop rotation. The lower level of crop rotation in medium-size farms still remains a question. Their incentives for catch crop cultivation and reasons for making choices need further investigation.

Eurostat (Statistical Office of the European Union) data from 2013 show that the average level of education in agriculture in Estonia is in line with the average level in Europe. This study illustrates that there is a strong correlation ($r = 0.79$) between the yield and the level of education (Fig. 10.4). The correlation is also affected by the farmers' age, which varies with different yields. Compared to other groups, the share of managers with basic education is the highest in Group 1, where the yields are the lowest, amounting to 25%. Around 12% of the managers have a secondary vocational or vocational education, which is one of the lowest indicators across the groups. Among the company executives, 25% have a secondary vocational or a vocational education in agriculture. The study also shows that 13% of the managers have a degree from university or from an applied science university, which is one of the highest across groups. 25% of the managers have an agricultural higher education or agricultural applied education. This proportion is one of the lowest compared to that in other groups.

In Group 2, 10% of the managers have a basic education. The share of managers with secondary vocational or vocational education is 38%, which is the highest across the groups, whereas the proportion of managers with agricultural secondary vocational and vocational education is 19%, which is one of the lowest. Of the managers, 33% have an agricultural higher education or higher applied education. Eighteen percent of the respondents belong to this group, whose average age is 51.6 years (Fig. 10.3).

Group 3 constitutes of 18% of the managers and their average age is 50 years (Fig. 10.3). Similar to the second group, 10% of the managers in the group have a

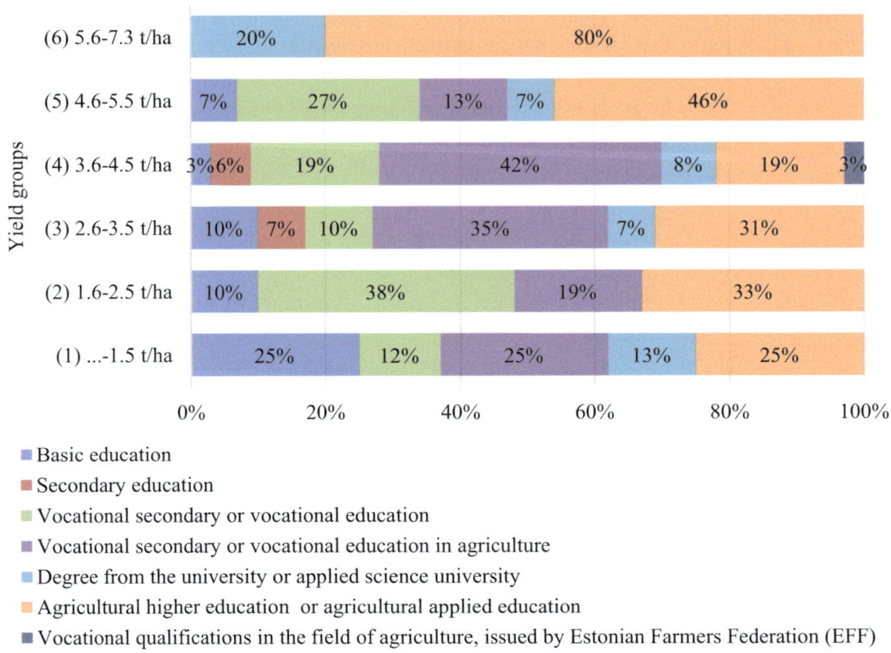

Fig. 10.4 Associating crop yields and levels of formal education of managers (Source: Aste 2016)

basic education (Fig. 10.4). The share of executives with non-agricultural secondary education and vocational secondary education is 7% and 10%, respectively. The proportion of managers with secondary or vocational education is the lowest. Thirty-five percent of the managers in this group have received an agricultural secondary or agricultural vocational secondary education, which is one of the highest indicators across the groups. Seven percent of the managers have acquired higher or applied higher education, and the share of agricultural higher education or agricultural higher vocational education constitutes 31%.

In group 4, the share of basic education is 3%, which is one of the lowest compared to other groups. Six percent of the managers have secondary education and 19% secondary vocational or vocational education. The group is characterized by the highest proportion of managers with agricultural secondary or agricultural vocational education, which is set at 42%. Eight percent have received higher or applied higher education and 19% agricultural higher education or agricultural higher vocational education. Only 3% of the managers in this group have vocational qualifications in the field of agriculture, issued by the Estonian Farmers Federation (EFF). To receive a professional qualification, an examination assessed by the professional qualifications committee must be taken (EPK 2014). Thirty-two percent of the entrepreneurs belong to this group, and their average age is 49.5 years (Fig. 10.3).

In group 5 the share of farmers with basic education amounts to 7% and with secondary or vocational secondary education to 27%, which is one of the highest results across vocational or vocational education (13%), and with higher education or applied higher education (7%), is the lowest among the groups; the group is in general engaged in sustainable intensification and the yields exceed the national average. This can be explained by the high proportion (46%) of managers with an agricultural higher education or professional higher education degree. The fifth group includes 13% of the managers, whose average age is 44.4 years (Fig. 10.3).

The yields are the largest in group 6 surpassing the national average almost twice. As many as 80% of the managers have acquired agricultural higher education or agricultural professional higher education, which is 1.7 times more than in the fifth group and 2–4 times more than in other groups (Fig. 10.4). Twenty percent of the managers have a higher education or professional higher education degree, which is the best result of all groups. Only 4% of the managers belong to this group and their average age is 42.8 years (Fig. 10.3).

10.4 Discussion

Results of the regression analysis did not meet the expectations of the researchers in all aspects. According to the literature, direct seeding should increase crop yields, and thus be one of the most important technological factors of sustainable intensification. The regression analysis, however, showed that direct seeding reduced spring wheat yields. In the case of conservation tillage, which is also considered a sustainable intensification technology, the results were as expected and showed in large yields. Ploughing is not regarded as a sustainable intensification technology and as demonstrated by the regression analysis, the application of this technology reduced the yields.

The application of seeds is in line with the common understanding, whereby higher quality seeds generate a positive effect on the yields. Purchased uncertified seeds reduced and purchased certified dressed seeds increased the yields for all three technologies applied. Own produced uncertified dressed seed increased and own produced uncertified undressed seed decreased the yields for all three technologies applied.

The use of fertilizers increased yields for all three technologies applied, which was to be expected. The data collected by this survey did not allow the examination of fertilization technologies used. Furthermore, data on the use of fertilization technologies would have made it possible to assess the efficiency of fertilization and its compliance with the concept of sustainable intensification. Organic liquid manure increased the yields across all technologies, and its effect on yields did not fall markedly behind that of the inorganic fertilizers on production.

The use of plant protection products increased the yields in the case of direct seeding only, which is not surprising given the specific nature of direct seeding where all crop residues remain on the surface of the arable land and therefore weeds are difficult to control.

The regression analysis suggests that conservation tillage is the best choice in terms of sustainable intensification and yields. The authors believe that it is advisable to use the highest quality seeds available. The expense of high-quality seed might be larger, but their effect on the yield is still considerably more. Mineral fertilizers must be used, but within the required standards and taking into account the needs of the specific crop. To ensure the efficient use of fertilizers, it is possible to introduce a site-specific technology that calculates the optimum amounts of fertilizer per each square metre of the field (precision fertilization). This is not the only solution, and ideally this technology should be used together with a variety of other effective techniques. As the effects of inorganic fertilizers on the outputs or yields do not differ greatly from those of organic fertilizers, greater use should be made of the latter to preserve the environment and maintain the humus balance of the soil. The use of plant protection products should be restricted, given that in most ecosystems their use is not necessary. As an alternative, pesticides can be substituted for catch crops and integrated pest management.

The analysis of average actual yields and potential yields across various technologies established that farms in the first (yields up to 1.5 t/ha), second (1.6–2.5 t/ha) and third (2.6–3.5 t/ha) groups showed a great potential for increasing the yields of spring wheat, which can be achieved with sustainable intensification technologies and inputs, complemented with the right approach. For example, despite the fact that ploughing is not considered to be a sustainable intensification technology, the larger than actually realised potential yields depend mainly on the use of other inputs of sustainable intensification. The technologies applied during ploughing also include elements of sustainable intensification and the information presented above reflects that in case of larger yields, high-quality inputs are used and their use is better targeted. Starting from the fourth group (yield 3.6–4.5 t ha^{-1}), the actual yield is more than the potential yields obtained through the application of technologies. This shows that the farms belonging to groups 4–6 already use better technology and the finest set of inputs. The authors consider that the given enterprises are already engaged in sustainable intensification. Therefore, the larger is the yield, the more likely it is that the company is committed to sustainable intensification. The companies in the first group should cooperate with the companies in group 5 (4.6–5.5 t ha^{-1}) and group 6 (5.6–7.3 t ha^{-1}) so that the producers could acquire the working practices and learn the sustainable use of inputs.

The first three groups are characterized by the fact that for the different technologies applied, the actual yields are smaller than the potential yields. For the next three groups, the situation is reversed — actual yields exceed the potential yields. The first three groups consisted of 61 companies, and the latter three of 58 companies. Thus, according to the present approach, it can be seen that 51.3% of the agricultural enterprises are not engaged in sustainable intensification against the 48.7% that are committed to it, which can be considered a good result. Still, the proportion of farms not involved in sustainable intensification could be smaller. For some enterprises, the use of proper and crop specific technologies, as well as the application of high-quality inputs could improve the situation.

The younger entrepreneurs are more committed to sustainable intensification. They make more active use of precision farming technologies that help them to achieve even higher yields. The analysis of the farmers' average age against the application of precision farming across different yields proved that the youngest farm managers are engaged in precision farming the most. In group 1 the average age was 51 years and the uptake of precision farming technologies was 37.5%, whereas in group 6 the same indicators were 43 years and 80%, which means that the activity index regarding precision farming of the farmers with the largest yields in group 6 is 42.5 percentage points more than for group 1. Based on the results of the analysis, the authors consider that the results confirm the theory presented: precision farming technologies can contribute to larger yields, and the use of precision farming indicates that sustainable intensification is practiced.

Age-related disincentives of sustainable intensification include conservatism and the inability and unwillingness of farmers to ask for help. Nowadays, farm managers have to be innovative and invest in the latest technology and high-quality inputs. In addition, educational institutions have an important role to play by introducing the latest tools, methods and solutions to the students of agriculture (including agribusiness). Younger entrepreneurs are more prone to take risks and to use the most recent, and perhaps not the most thoroughly tested technologies that facilitate larger yields and raise the level of sustainable intensification of the farm.

In the analysis of the application of catch crops against farms with different yields, there was no correlation between intercropping, sustainable intensification and yields. The reasons for this may lie in the specifics of the technology used or possibly in the lack of the necessary technology, which create suitable conditions for growing homogeneous crops. Another reason may be the sequence and choice of crops in crop rotation, which does not include a lot of cereals. One of the causes may come from the higher market prices of the crops cultivated in the given years because farmers prefer to produce crops that entail lower risks and promise higher returns, regardless of the need to grow catch crops. Only a few farmers canvassed used catch cropping in the crop rotation system. However, their application in crop rotation provides the producers with another opportunity to move towards sustainable intensification by increasing yields and reducing negative environmental impacts.

When analysing the correlation between the farm managers' level of education and yields, it emerged that the share of managers with basic education was the highest in the lowest-yield group. Group 2 was characterized by a large share of farmers with secondary vocational or vocational education (38%), and agricultural higher education or higher applied education (33%). To achieve higher yields through sustainable intensification, it is not enough to have practical knowledge, but it is also necessary to develop the theoretical approach and a thorough knowledge of modern agriculture. In the third group, 35% of the managers have agricultural secondary vocational or agricultural vocational education, which is one of the highest figures in the groups. Seven percent of the managers in this group had acquired higher education or applied higher education. Although the acquired tertiary education does not provide the necessary practical agricultural knowledge, the graduates of higher education establishments have definite advantages over the graduates of secondary, secondary vocational and vocational education because higher education grants a broader knowledge and skills set necessary for finding and processing information. To achieve the maximum yields through sustainable intensification, the knowledge and skills must be constantly upgraded over time, which means that agricultural producers must receive support from the contemporary agricultural extension system or advisory services.

Group 4, where the producers are already to a certain extent engaged in sustainable intensification, is characterized by a relatively high percentage of secondary vocational or vocational school graduates (42%). In group 5, which is quite effectively involved with sustainable intensification (the yields exceed the national average), the share of managers with agricultural higher or agricultural professional higher education is significantly higher – 46%. The value of this indicator is higher only in group 6 where the yields are the largest and exceed the national average almost twice. This group stands out for significant sustainable intensification. If various levels of education are represented in other groups, all managers in this group have a tertiary degree. University graduates have substantial knowledge and understanding, which creates a good basis for innovative and effective farming. The authors find that agricultural higher education prepares the ground for sustainable intensification. The managers in the fifth and sixth groups are consciously dealing with decreasing the adverse environmental impact while aiming for higher yields.

The study revealed that the factors affecting sustainable intensification recognised in the scientific literature are similar to those encountered on Estonian cereal farms. To assess the level of sustainable intensification of enterprises, it is possible to proceed from the level of education, age, application of precision farming technology to the use of catch crops and yields. The analysis of the above-mentioned factors suggests that nearly half of the companies in Estonia are engaged in sustainable intensification. In the companies not yet engaged in sustainable intensification, it is not necessary to make radical changes in the production process. Transition to sustainable intensification may be made step by step, assessing the impact of the decisions on the results along the way.

10.5 Conclusions

The present research analysed the level of sustainable intensification among Estonian crop production enterprises. Potential yields of spring wheat were calculated separately within each technology using a variety of inputs that are the indicators of sustainable intensification. The results suggest that 51.3% of the enterprises are not engaged in sustainable intensification, which is a very high proportion in the light of the positive impact that sustainable intensification could have on Estonian agriculture. It turned out that among the entrepreneurs, 48.7% adhere to the principles of sustainable intensification and engage in attendant technology and processes. This is reflected in the fact that the actual yields on the farms are higher than the potential yields calculated on the basis of the technologies (with inputs) selected by the authors.

The analysis confirmed that the level of sustainable intensification and crop yield is related to the application of precision farming technologies and the manager's age. More than 60% of the farms practicing sustainable intensification use precision farming. For those agricultural enterprises that are not engaged in sustainable intensification, the yields were less than half of those practising sustainable intensification. These differences may be explained by a variety of reasons, but the economic capacity of the farm and the area of arable land are among them. Based on the results, it can be argued that precision farming could be an important indicator for the achievement and assessment of the level of sustainable intensification.

Research into the educational level of business managers in relation to yields revealed that the average age of the managers involved in sustainable intensification is under 50 years, whereas the average age of those not engaged in sustainable intensification was over 50. The advanced age of farm managers poses a challenge to Estonian agriculture, which in some cases is exacerbated by the lack of successors to present farmers. The authors, however, believe that the situation is changing for the better because agriculture as an economic sector is gaining popularity and an increasing number of young people are applying for a study place in agriculture-related specialities in Estonia.

The analysis did not provide a clear picture of the effect of catch crops on yields. It transpired that out of the surveyed enterprises, the farms not engaged in sustainable intensification made more use of intercropping. Sustainable intensification does depend on the use of catch crops because in the sixth group 40% of the farmers practiced intercropping.

The level of education is strongly correlated with the level of sustainable intensification. A higher agricultural education establishes a sound basis for sustainable intensification, which is targeted at achieving higher yields and the minimization of adverse environmental impacts.

The present research allows us to conclude that it is essential to increase cooperation between the enterprises operating in the agricultural sector. Farms practicing sustainable intensification should be brought together with farms not involved in it. More emphasis should be put on the application of modern ICT opportunities in agriculture and business management (Big Data). Thus, the employees and managers could acquire theoretical and practical knowledge about how to use sustainable intensification technologies in their production processes. Company executives or employees responsible for production on a daily basis should be solution-oriented and keep up with the professional developments, changes in global agriculture and agricultural innovation.

The present study concentrated mostly on the analysis of social factors, technology adoption and yields in the light of sustainable intensification, leaving the economic aspects of sustainable intensification for further research.

The concept of sustainable intensification and the direct economic benefits needs to be studied further in agriculture. It is necessary to raise the awareness of agricultural producers about the possibilities to increase productivity in a sustainable way. Aside from economic aspects of increased productivity, more attention needs to be paid to environmental friendly pathways and to enable additional training in the agricultural sector.

Appendices

Appendix A: Indicators of Sustainable Intensification Used in the Correlation Analysis

x_1 ID of the business enterprise,
x_2 direct seeding technology,
x_3 conservation tillage,
x_4 ploughing,
x_5 spring wheat yield,
x_6 purchased seed,
x_7 own produced seed,
x_8 purchased certified seed,
x_9 purchased uncertified seed,
x_{10} purchased certified dressed seed,
x_{11} purchased certified undressed seed,
x_{12} purchased uncertified dressed seed,
x_{13} purchased uncertified undressed seed,
x_{14} purchased dressed seed,
x_{15} purchased undressed seed,
x_{16} own produced certified seed,
x_{17} own produced uncertified seed,
x_{18} own produced certified dressed seed,
x_{19} own produced certified undressed seed,
x_{20} own produced uncertified dressed seed,
x_{21} own produced uncertified undressed seed,
x_{22} own produced dressed seed,
x_{23} own produced undressed seed,
x_{24} plant protection products were used,
x_{25} plant protection products were not used,
x_{26} fertilizers were used,
x_{27} fertilizers were not used,
x_{28} liquid fertilizers,
x_{29} granular fertilizers,
x_{30} organic liquid manure was used,
x_{31} organic liquid manure was not used,
x_{32} organic solid manure was used,
x_{33} organic solid manure was not used.

Appendix B: The Selected Variables for Regression Analysis

The selected variables for regression analysis, where:

x_2 direct seeding,

x_3 conservation tillage,
x_4 ploughing,
x_5 spring wheat yields,
x_9 purchased uncertified seed,
x_{10} purchased certifies dressed seed,
x_{20} own produced uncertified dressed seed,
x_{21} own produced uncertified undressed seed,
x_{26} fertilizers were used,
x_{24} plant protection products were used,
x_{30} organic liquid manure was used.

Appendix C: Direct Seeding, Conservation Tillage, Ploughing Technologies Regression Analysis Parameters

	Direct seeding, regression coefficient	Conservation tillage, regression coefficient	Ploughing, regression coefficient
x_2 direct seeding	−0.2214	–	–
x_3 conservation tillage	–	0.3242	–
x_4 ploughing	–	–	−0.2110
x_9 purchased uncertified seed	−0.8543*	−0.7536	−0.7785
x_{10} purchased certifies dressed seed	0.3678	0.3688	0.3900
x_{20} own produced uncertified dressed seed	0.3605	0.3337	0.3567
x_{21} own produced uncertified undressed seed	−0.8675*	−0.7992*	−0.8352*
x_{26} fertilizers were used	1.1462*	1.3178*	1.2553*
x_{24} plant protection products were used	0.0450	−0.1177	−0.0809
x_{30} organic liquid manure was used	0.8966*	0.8397	0.8489
Constant	2.2516*	2.0594*	2.3052*
n	119	119	119
R^2	0.3254	0.3366	0.3280
Sign F	4.7155E-07	2.0720E-07	3.9083E-07

Source: Compiled by the authors based on data Eesti Maaülikool 2012
n number of observations. * statistically significant

References

AFI (2013) Sustainable intensification: a new paradigm for African agriculture. A 2013 Montpellier Panel Report. Agriculture for impact (AFI). Imperial College London. http://ag4impact.org/wp-content/uploads/2013/04/MP_0176_Report_Redesign_2016.pdf. Accessed 5 Sept 2016

Ariva J, Viira A-H, Põldaru R, Roots J (2015) Medium-run projections for greenhouse gas emissions arising from agriculture: the case of milk production in Estonia. Agric Food Sci 24(4):300–312

Aste R (2016) Sustainable intensification theoretical starting points in agriculture and practice in crop sector so far example of a 2012 case study. Bachelor's thesis. Estonian University of Life Sciences, Institute of Economics and Social Sciences, Tartu. (in Estonian)

Barnes AP (2012) Sustainable intensification in Scotland: a discussion document. Rural Policy Centre, SAC, Edinburgh (March). www.sruc.ac.uk/download/downloads/id/472/sustainable_intensification_in_scotland. Accessed 2 Sept 2016

Barnes AP, Poole CEZ (2012) Applying the concept of sustainable intensification to Scottish agriculture. Contributed Paper prepared for presentation at the 86th annual conference of the agricultural economics society, University of Warwick, United Kingdom, 16–18 April 2012. http://ageconsearch.umn.edu/bitstream/134710/2/Andrew_Barnes_Barnes_Poole_AES.pdf

Barnes AP, Thomson SG (2014) Measuring progress towards sustainable intensification: how far can secondary data go? Ecol Indic 36:213–220

Buckwell A, Uhre AN, Williams A et al (2014) Sustainable intensification of European agriculture. A review sponsored by the RISE Foundation. Public Utility Foundation for Rural Investment Support for Europe (RISE), Brussels. http://www.ieep.eu/assets/1404/111120_BROCH_SUST_INTENS_DEF.pdf. Accessed 31 Aug 2016

Campbell BM, Thornton P, Zougmoré R et al (2014) Sustainable intensification: what is its role in climate smart agriculture? Curr Opin Environ Sustain 8:39–43

Cook S, Silici L, Adolph B et al (2015) Sustainable intensification revisited. IIED Issue Paper. IIED, London. http://pubs.iied.org/pdfs/14651IIED.pdf. Accessed 31 Aug 2016

de Haan C, Setshwaelo LL (2015) An environmental perspective with particular attention to Animal genetic resources. Sustainable intensification of livestock production systems. http://agrienvarchive.ca/bioenergy/download/deHaan_sus_intens_livestockprod.pdf. Accessed 8 Sept 2016

Dillon EJ, Hennessy T, Buckley C et al (2014) The sustainable intensification of the Irish Dairy Sector. Contributed paper prepared for presentation at the 88th annual conference of the agricultural economics society, AgroParisTech, Paris, France, 9–11 April 2014. http://ageconsearch.umn.edu/bitstream/169740/2/Emma%20Jane_Dillon_AES%20Sustainability%20Dairy_Dillon%20et%20al.-1.pdf. Accessed 1 Sept 2016

Drechsel P, Heffer P, Magen H et al (eds) (2015) Managing water and fertilizer for sustainable agricultural intensification, 1st edn. International Fertilizer Industry Association (IFA), International Water Management Institute (IWMI), International Plant Nutrition Institute (IPNI), and International Potash Institute (IPI), Paris, France

Elliott J, Firbank LG, Drake B et al (2013) Exploring the consept of sustainable intensification. report. Land Use Policy Group, Edinburgh. http://www.snh.gov.uk/docs/A931058.pdf. Accessed 19 Mar 2016

EPK (2014) Põllumajandusvaldkonna kutsed. – Eesti Põllumeeste Keskliit (EPK). http://eptk.ee/kutse-andmine/kutseandmise-kord/. Accessed 24 Sept 2016 (in Estonian)

FAO (2011) Save and grow. A policymaker's guide to the sustainable intensification of smallholder crop production. Food and Agriculture Organization of the United Nations (FAO), Rome. http://www.fao.org/3/ea20b790-4a84-57fc-aa98-a367a012ddb8/i2215e.pdf. Accessed 31 Jan 2017

FAO (2014) Building a common vision for sustainable food and agriculture. Principles and approaches. Food and Agriculture Organization of the United Nations (FAO), Rome. http://www.fao.org/3/a-i3940e.pdf. Accessed 31 Aug 2016

Garnett T, Godfray HCJ (2012) Sustainable intensification in agriculture. Navigating a course through competing food system priorities, Food Climate Research Network and the Oxford Martin Programme on the Future of Food. University of Oxford, Oxford. http://www.fcrn.org.uk/sites/default/files/SI_report_final.pdf. Accessed 31 Aug 2016

Hogan P (2015) Speech by Commissioner Phil Hogan at ICT-agriculture conference. Delivering sustainable intensification through collaboration, Ireland. http://ec.europa.eu/agriculture/commissioner-speeches/pdf/hogan-kilkenny-01-05-2015_en.pdf. Accessed 7 Sept 2016

Huggett G (2015) Sustainable intensification research platform. In: Devon farming and environment 2015 conference, Newton St Cyres, England. http://www.fwagsw.org.uk/files/1514/2503/7260/Exeter_Uni_Sustainable_Intensification_Research.pdf. Accessed 19 Mar 2016

Kassam A, Friedrich T, Shaxson F et al (2011) Production systems for sustainable intensification integrating productivity with ecosystem services. Technikfolgenabschätzung, Theor. Prax. 20(2):38–45

Lampkin NH, Pearce BD, Leake AR et al (2015) The role of agroecology in sustainable intensification. Report for the land use policy group. Organic Research Centre, Elm Farm and Game & Wildlife Conservation Trust. http://www.snh.gov.uk/docs/A1652615.pdf. Accessed 19 Mar 2016

Loos J, Abson DJ, Chappell MJ et al (2014) Putting meaning back into "sustainable intensification". Front Ecol Environ 12(6):356–361

Pretty J, Bharucha ZP (2015) Integrated pest management for sustainable intensification of agriculture in Asia and Africa. Insects 6(1):152–182

Schiefer J, Lair GJ, Blum WEH (2015) Indicators for the definition of land quality as a basis for the sustainable intensification of agricultural production. Int Soil Water Conserv Res 3(1):42–49

SDSN (2013) Solutions for sustainable agriculture and food systems. Technical report for the post-2015 development agenda. Sustainable Development Solutions Network (SDSN). http://unsdsn.org/wp-content/uploads/2014/02/130919-TG07-Agriculture-Report-WEB.pdf. Accessed 1 Sept 2016

Statistic Estonia (SE) (2016) Statistics Estonia (SE). Online database. http://www.stat.ee/en. Accessed 8 Sept 2016

Smith A, Thorne P, Snapp S (2015) Measuring sustainable intensification in smallholder agroecosystems: a review. International Livestock Research Institute. https://cgspace.cgiar.org/bitstream/handle/10568/67181/AR_SI_metrics_mar2015.pdf?sequence=1. Accessed 1 Sept 2016

TRS (2009) Reaping the benefits: science and the sustainable intensification of global agriculture. The Royal Society. https://royalsociety.org/~/media/Royal_Society_Content/policy/publications/2009/4294967719.pdf. Accessed 31 Aug 2016

TWB (2005) Agriculture investment sourcebook. Agriculture and rural development. The World Bank (TWB). http://www-wds.worldbank.org/external/default/WDSContentServer/WDSP/IB/2005/11/18/000011823_20051118163155/Rendered/PDF/343920PAPER0Ag101OFFICIAL0USE0ONLY1.pdf. Accessed 20 Mar 2016

USG (2015) Feed the future. Program for sustainable intensification. U.S. Government. https://www.feedthefuture.gov/sites/default/files/resource/files/ftf_factsheet_fsicsustainableint_may2015.pdf. Accessed 31 Aug 2016

Chapter 11
How to Model the Adoption and Perception of Precision Agriculture Technologies

Giacomo Carli, Vilma Xhakollari, and Maria Rita Tagliaventi

Abstract The adoption of precision agriculture has shown to positively affect the performance of farms, even though its benefits vary according to the size of farms and their location. In light of the promising avenue that precision agriculture opens up, it is essential to understand which factors may facilitate its diffusion, and through which processes. This chapter focuses on the models proposed to explain technology adoption: Theory of Reasoned Action, Theory of Planned Behaviour, Motivational Model, Technology Acceptance Model, TAM2 and TAM3, Combined TAM and TPB, Model of PC Utilization, Innovation Diffusion Theory, Social Cognitive Theory and Unified Theory of Acceptance and Use of Technology. We analyse contributions targeting specifically the agricultural domain. Remarkably, most models and papers share the perspective that individual factors account for the willingness of individuals to engage in technology adoption, and there is a progressive commonality of factors between models based on different theories. In addition to individual-level features, some models analyse the relevance of environmental and social factors in prompting technology diffusion, thus depicting a more comprehensive framework to aid understanding of the dynamics linked to the adoption of precision agriculture. Eventually, some reflection on how to expand knowledge of precision agriculture along this line of reasoning aimed at integrating personal and social characteristics is offered. The importance of social network patterns and of social support in entrepreneurial initiatives that sustain adoption of precision agriculture is stressed in this chapter.

G. Carli (✉)
Department of Strategy and Marketing, The Open University Business School, Milton Keynes, UK
e-mail: Giacomo.Carli@open.ac.uk

V. Xhakollari
Department of Agricultural Sciences, University of Bologna, Bologna, Italy
e-mail: vilma.xhakollari2@unibo.it

M.R. Tagliaventi
Department of Management, University of Bologna, Bologna, Italy
e-mail: maria.tagliaventi@unibo.it

S.M. Pedersen, K.M. Lind (eds.), *Precision Agriculture: Technology and Economic Perspectives*, Progress in Precision Agriculture,
https://doi.org/10.1007/978-3-319-68715-5_11

Keywords Precision agriculture • Technology adoption • Behavioural models • Intention to adopt • Technology acceptance model • Factor analysis • Reliability

11.1 Introduction

A recent survey in the USA shows that Precision Agriculture (PA) adopters can increase their profits by $66 per acre (Schimmelpfennig 2016). Nevertheless, this value is strongly affected by the size of the farms, with higher benefits for larger farms because of economies of scale. European farms are considerably smaller than USA farms with an average of 175 ha in the USA compared to just 16 ha in Europe (Census of Agriculture 2012; Eurostat Farm structure statistics 2013), and size is believed to be one of the causes of a low diffusion of PA technologies in Europe.

It is, however, important to observe that this structural difference not only affects the availability of financial resources for farmers to fund the adoption of site-specific solutions, but it might affect the whole process of adoption. The fact that European farmers are mainly smallholders effects the adoption of technology and makes the process based on adoption beyond purely rational evaluations. Moreover, while the 70–80% of new farm equipment is manufactured with some kind of PA technology embedded, the advantages stemming from PA in Europe remain limited and can be increased only by complementing different technologies in a more complex, yet expensive system (Zarco-Tejada et al. 2014). We argue that farmers' evaluations might not be based solely on the net benefits of the investment in technology, but also on perceptions related to innovation and social factors. Accordingly, it is important to improve our understanding of how and why farmers come to the decision to adopt PA technologies. To delve into this issue, we look at studies on how potential adopters decide on the use of new innovations. We conducted a careful review of the approaches and of the contributions in the so-called ex-ante studies (Pierpaoli et al. 2013; Pignatti et al. 2015). Our review depicts the state-of-art of the research on the adoption of agriculture-related technology to aid the design of new studies on the topic.

In the first section, we introduce the theoretical models that can be applied to evaluate the adoption of technological innovation in different fields. We reviewed nine models: Theory of Reasoned Action, the Theory of Planned Behaviour, the Motivational Model, the Technology Acceptance Model, the TAM2 and TAM3, the Combined TAM and TPB, the Model of PC Utilization, the Innovation Diffusion Theory, Social Cognitive Theory, and the Unified Theory of Acceptance and Use of Technology. In drawing our comparison, we depict which theoretical constructs are typically measured in the models to show areas of overlap across different models.

In the second paragraph, we focus on the domain of agriculture looking at how the models introduced in the previous paragraph have been applied in this field. We conduct a review on ex-ante evaluation of innovations in the broad agricultural domain. We delve into 16 papers and compare their findings to elucidate the current understanding of adoption of agriculture-related technology. The comparison of models proposed in the 16 papers shows how different factors can be introduced to explain the decision to adopt. As models are oriented to overlap and merge progressively, the analysis depicts areas for further development of the models.

In the last paragraph, we discuss possible avenues of future development of the analysis of the ex-ante adoption models to account for the dual nature of this process—individual and social. The discussion entails the importance of evaluating the interrelation of individual and social aspects and the role of institutional actors as possible promoters of the adoption of agricultural technologies.

11.2 Theoretical Models

The aim of this section is to introduce the principal models used in the evaluation of technology adoption. We found nine models applied in the evaluation of technology adoption, and in this paragraph we present their rationale and the dimensions considered as affecting technology adoption.

The Theory of Reasoned Action (TRA) and the Theory of Planned Behaviour (TPB) are the earliest theories; they derive from social psychology and are used to explain the use of Information Systems (IS). At that stage, it was seen as relevant to understand first the generic underlying behaviour related to the adoption of new technologies, and then to proceed with specific models focused entirely on understanding and identifying the behavioural factors affecting IS usage.

The **Theory of Reasoned Action (TRA)** was formulated in social psychology and was originally developed by Fishbein in 1967. It aims to explain human behaviour; the TRA maintains that behaviour is controlled by intention and the stronger the intention is, the harder the effort to perform the action will be (Fishbein 1967; Fishbein and Ajzen 1975; Ajzen and Fishbein 1980). According to Fishbein and Ajzen (1975), two factors determine the intention: the Attitude toward the Behaviour and the Subjective Norms. The former is affected by the beliefs on the outcome of the behaviour and by the individual evaluation of that outcome, either positive or negative. The latter is related to individual perceptions on what society thinks of the behaviour. Subsequent research has found that Experience and Voluntariness are two relevant factors in explaining a behaviour (Karahanna et al. 1999): with the increase in experience, Attitude towards the Behaviour becomes more important, while the relevance of Subjective Norms diminishes. However, Hartwick and Barki (1994) showed that when users are not forced to adopt technology, Subjective Norms become more important supporting the inclusion of Voluntariness in the model.

The **Theory of Planned Behaviour (TPB)** is an extension of the TRA (Ajzen and Fishbein 1980). According to Sheppard et al. (1988), the theory explains a broader range of behaviour in comparison to TRA. This is because TPB, besides attitudes and subjective norms, comprises a third factor: perceived behavioural control (PBC). This latter consists of "the perceived ease of use or difficulty of performing the behaviour" (Ajzen 1991).

The TPB has been applied successfully to analyse users' behaviour regarding different types of technologies (Harrison et al. 1997; Mathieson 1991). In this context, the Perceived Behavioural Control is related to the perceptions of internal and external constraints on behaviour (Davis et al. 1989; Taylor and Todd 1995b).

Similarly to TRA, in TPB Experience and Voluntariness were not included in the original model. As shown for TRA, however, research has found these two factors are important in explaining the behaviour *vis-à-vis* technology adoption (Venkatesh and Morris 2000; Karahanna et al. 1999).

Regarding gender, research has shown that Attitude was more relevant for men, whereas Subjective Norms and PBC were found to affect both men and women with limited experience (Venkatesh et al. 2000).

Finally, even though age was not included in the original model, Morris and Venkatesh (2000) concluded that Subjective Norms were more relevant to older women. On the other hand, Perceived Behavioural Control was more relevant for seasoned workers, whereas Attitude was more important for younger workers.

The **Motivational Model (MM)** ensued from the Motivational Theory (Davis et al. 1992). According to Deci and Ryan (1985), motivation is the most important factor that affects behaviour in different fields. Drawing on a wide variety of studies, research has grouped motivational factors into two main categories: intrinsic and extrinsic factors (Deci 1971; Deci and Ryan 1985): "*Intrinsic motivation refers to the pleasure and inherent satisfaction derived from a specific activity*" (Venkatesh and Speier 1999, p. 2; Deci 1975; Vallerand 1997). Examples of intrinsic motivation include Enjoyment and Playfulness. Conversely, *"extrinsic motivation emphasizes performing a behaviour because it is perceived to be instrumental in achieving valued outcomes that are distinct from the activity such as increased pay and improved job performance"* (Venkatesh and Speier 1999, p. 2; Lawler and Porter 1967; Vroom 1964). Perceived Usefulness, Perceived Ease of Use and Subjective Norms are examples of extrinsic motivation.

Davis et al. (1992) applied the theory to the domain of technology usage. They found that office workers' intention to use computers depends primarily on their perceptions of how computer-usage would improve their work performance (Usefulness), and secondly, by the enjoyment they experience while using the computers (Enjoyment). A particularly striking result was that when determining intentions, Usefulness is four to five times more influential than enjoyment.

Thus, considering the motivational theory and the findings from Davis et al. (1992), extrinsic motivations are far more important and influential when deciding about technology usage.

The **Technology Acceptance Model (TAM)** was adapted from the TRA (Ajzen and Fishbein 1980; Fishbein and Ajzen 1975) to depict the factors affecting technology usage. In the TAM, the intention to adopt technology is determined by two principal factors: Perceived Usefulness, i.e. "*the degree to which a person believes that using particular system would enhance his/her job*" (Davis 1989, p. 320); and Perceived Ease of Use, i.e. *"the degree to which a person believes that using a particular system would be free of effort"* (Davis 1989, p. 320).

Studies have concluded that TAM explains approximately 40% of the total variance of behaviour and intention to use technology (Venkatesh and Davis 2000).

TAM has been refined over the years and been later elaborated as **TAM2**. This extended version of TAM is derived from the TRA and the TPB described above. Additional factors related to social influence processes (Subjective Norm, Voluntariness, and Image[1]) and cognitive instrumental processes (Job Relevance, Output Quality, Result Demonstrability, and Perceived Ease of Use) are incorporated into the TAM2 (Venkatesh and Davis 2000).

Later on, Venkatesh and Bala (2008) proposed another extended version of TAM2, which has been labelled as **TAM3**. This new evolution introduces experience as a moderating factor of the relation between three couples of factors: (1) Perceived Ease of Use and Perceived Usefulness; (2) Computer Anxiety and Perceived Ease of Use; (3) Perceived Ease of Use and Behavioural Intention to Adopt.

Gender and age were not included in either of the three versions of the TAM. Nevertheless, Venkatesh and Morris (2000) found that perceived usefulness was more relevant for men than women, whereas ease of use was more important for women than for men.

Taylor and Todd (1995a) introduced a new model, which consists of a combination of the two previous models, TAM and TPB, and has been called "The **Combined TAM and TPB (C-TAM-TPB)**". The new model comprises factors emerging from the two original models: Attitude toward Behaviour, Subjective Norm, Perceived Behavioural Control and Perceived Usefulness. Taylor and Todd (1995a) argued that this model can be applied to both experienced and inexperienced users. According to the same study, for both groups of individuals, all other determinants, except for Attitude, were significant. Thus, this model version might be used successfully to predict the behaviour prior to the implementation of a technology.

The **Model of PC Utilization (MPCU)** was elaborated by Triandis in 1980 to allow for a better understanding of the determinants of behaviour introduced in the TRA. According to this theory, *"behavioural intentions are determined by feelings people have toward the behaviour (affect), what they think they should do (social factors), and by the expected consequences of the behaviour"* (Thompson et al. 1991, p. 125). In other words, behaviour is affected by habits, intentions and facilitating conditions. Thompson et al. (1991) included this theory into the IS context to predict PC usage in the workplace. Venkatesh et al. (2003) have later considered the model in their research on the unification of technology acceptance models, but unlike Thompson et al. (1991) and in line with Triandis (1980), they focused on intention rather than on behaviour.

The following factors are considered in the model:

- Job Relevance – *"the extent to which an individual believes that using a technology can enhance the performance of his or her job"* (Thompson et al. 1991, p. 129).

---

[1] The degree to which an individual perceives that use of an innovation will enhance his or her status in his or her social system (Moore and Benbasat 1991, p. 195).

- Complexity – *"The degree to which an innovation is perceived as relatively difficult to understand and use"* (Rogers and Shoemaker 1971; p.154).
- Long term Consequences – *"Outcomes that have a pay-off in the future"* (Thompson et al. 1991, p. 129).
- Affect towards Use – Based on Triandis (1980), affect toward use is a *"feeling of joy, elation, or pleasure, or depression, disgust, displeasure, or hate associated by an individual with a particular act"* (Triandis 1980, p. 211).
- Social Factors – *"the individual's internalization of the reference group's subjective culture, and specific interpersonal agreements that the individual has made with others, in specific social situations"* (Triandis 1980, p. 210).
- Facilitating Conditions – *"objective factors, 'out there' in the environment that several judges or observers can agree make an act easy to do"* (Triandis 1980, p. 205. In the Information Systems context, "provision of support for users of PCs may be one type of facilitating condition that can influence system utilization" (Thompson et al. 1991).

Remarkably, the **Innovation Diffusion Theory (IDT)** has been used to study a broad range of innovations, from agricultural tools to industrial technologies (Tornatzky and Klein 1982). Moore and Benbasat (1991) have adapted the model to technology acceptance, modifying the characteristics defined by Rogers (1995) and refining a set of constructs. The factors that are taken into account in the model are as follows:

- Relative Advantage – *"the degree to which an innovation is perceived as being better than its precursor"* (Moore and Benbasat 1991, p. 195).
- Perceived Ease of Use – *"the degree to which an innovation is perceived as being difficult to use*" (Moore and Benbasat 1991, p. 195).
- Image – *"the degree to which use of innovation is perceived to enhance one's image or status in one's social system"* (Moore and Benbasat 1991, p. 195).
- Visibility – *"the degree to which one can see others using the system in the organization"* (Venkatesh et al. 2003, p. 431; Moore and Benbasat 1991).
- Compatibility – *"the degree to which an innovation is perceived as being consistent with the existing values, needs, and past experiences of potential adopters"* (Rogers 2003, p. 15; Moore and Benbasat 1991, p. 195).
- Results Demonstrability – *"the tangibility of the results of using the innovation, including their observability and communicability"* (Moore and Benbasat 1991, p. 203).
- Voluntariness of Use – *"the degree to which use of the innovation is perceived as being voluntary, or of free will"* (Moore and Benbasat 1991, p. 195).

A different view on models addressing technology adoption has been inspired by the **Social Cognitive Theory (SCT)**. This theory holds that individuals' knowledge is related to the information they obtain by other subjects who perform a behaviour (Bandura 1986). Compeau and Higgins (1995) extended the original model to target the context of computer usage. Their formulation made possible application of the SCT model to the evaluation of technology acceptance.

Factors included in the model are:

- Output Quality – "*performance expectations deal with job-related outcomes*" (Venkatesh et al. 2003, p. 432; Compeau and Higgins 1995).
- Outcome Expectations – personal – "*it deals with individuals' esteem and sense of accomplishment*" (Venkatesh et al. 2003, p. 432; Compeau and Higgins 1995).
- Self-Efficacy – "judgement of one's ability to use a technology" (Venkatesh et al. 2003, p. 432).
- Perceived Enjoyment – "an individual's liking for a particular behaviour" (Venkatesh et al. 2003, p. 432).
- Computer Anxiety – "*evoking anxious or emotional reactions when it comes to performing the behaviour*" (Venkatesh et al. 2003, p. 432).

Finally, the so-called **Unified Theory of Acceptance and Use of Technology (UTAUT)** is a model developed through the integration and further development of the eight models described above. Targeting users' intention to include IT systems in their daily work, Venkatesh et al. (2003) showed that the eight models accounted for between 17 and 53% of the variation. Performance Expectancy, Effort Expectancy, Social Influence, and Facilitating Conditions are considered the main determinants of Behaviour, whereas Gender, Age, Experience and Voluntariness of use are considered as moderating factors (Venkatesh et al. 2003).

Table 11.1 reports the factors explaining the behaviour of all the above-mentioned models. As seen, TAM is the model that includes most of the factors that explain technology adoption behaviour.

Voluntariness and experience are considered the most relevant factors affecting the behaviour. This is because they emerge as explanatory variables in most of the models taken into consideration in this review.

It is important to stress that all the theories concerned in explaining the users' behaviour for IS usage are affected or based on the TRA and the TPB. Nevertheless, recent research has further developed and improved those theories, and the UTAUT is an example of the most recent theory regarding IS usage.

Table 11.1 represents a summary of the above mentioned behavioural models and the factors from which the behaviour of adopting a new technology depends. The TAM3 is the model that accounts for most of the factors. The TPB and IDT models depend on eight factors, which is half of the factors of TAM3.

The next paragraph will introduce some applications of these models and theories to the agricultural sector.

Table 11.1 Factors used in the models related to technology adoption behaviour

Factors	Models								
	Theory of Reasoned Action (TRA)	Theory of Planned behaviour (TPB)	Motivational Model (MM)	Technology Acceptance Model (TAM3)	Combined TAM and TPB (C-TAM. TPB)	Model of PC Utilization (MPCU)	Innovation Diffusion Theory (IDT)	Social Cognitive Theory (SCT)	Unified Theory of Acceptance and Use of Technology (UTAUT)
Attitude toward Behaviour (AB)	✓	✓			✓				
Subjective Norm (SN)	✓	✓		✓	✓				
Perceived Usefulness (PU)				✓	✓				
Perceived Ease of Use (PEOU)				✓			✓		✓
Computer Self-Efficacy (CSE)				✓					
Perceptions of Internal Control (PIC)		✓			✓				
Perceptions of External control (PEC)		✓		✓	✓				
Computer Playfulness (CPLAY)				✓					
Computer Anxiety (CANX)				✓				✓	

Perceived Enjoyment (ENJ)				✓				✓	
Voluntariness (VOL)	✓	✓		✓	✓		✓		
Image (IMG)				✓			✓		
Job Relevance (REL)				✓		✓			
Output Quality (OUT)				✓				✓	✓
Result Demonstrability (RES)				✓			✓		
USE (USE)				✓				✓	
Extrinsic Motivation (EM)			✓						
Intrinsic Motivation (IM)			✓						
Complexity (CPX)						✓			
Long Term Consequences (LTC)						✓			
Affect Towards Use (ATU)						✓			
Social Factors (SC)						✓			✓
Facilitating Conditions (FC)						✓			✓
Relative Advantage (RA)							✓		
Visibility (VIB)							✓		

(continued)

Table 11.1 (continued)

Factors	Models								
	Theory of Reasoned Action (TRA)	Theory of Planned behaviour (TPB)	Motivational Model (MM)	Technology Acceptance Model (TAM3)	Combined TAM and TPB (C-TAM. TPB)	Model of PC Utilization (MPCU)	Innovation Diffusion Theory (IDT)	Social Cognitive Theory (SCT)	Unified Theory of Acceptance and Use of Technology (UTAUT)
Compatibility (COB)							✓		
Outcome Expectations Personal (OEP)								✓	
Experience	✓	✓		✓		✓	✓		✓
Gender		✓		✓					✓
Age		✓							✓

11.3 Behavioural Models and Their Application in Agricultural Sciences

After introducing a detailed picture of behavioural models about technology adoption, this section shows how they have been applied to study the adoption of agriculture-oriented technologies.

To conduct a careful review of the research studies available on the topic, we resorted to Scopus and Google Scholar to search for articles using keywords related to the specific sector, such as "agriculture", "farm", "food production", in combination with keywords related to technology adoption, for instance "technology adoption", "behavioural models" and "technology acceptance". The numerous results were then divided into two groups: the empirical studies related to the behavioural models introduced in Paragraph 1, and the studies with an ex-post approach regarding technology adoption. We focused on the contributions analysing farmers and specialists' behaviour prior to the choice of adoption.

We identified sixteen papers: eight studies are based on the technology acceptance model, six on the theory of planned behaviour, and the remaining two studies combine the Theory of Reasoned Action, the Model of PC Utilization and the Innovation Diffusion Theory.

We analysed each paper to identify its theoretical underpinnings, the method and the setting studied. We summarize the main findings of the sixteen papers in Table 11.2.

Articles included in this review gauge attitudes and intentions to adopt innovative technologies in farm activities. As previously mentioned, this review has taken an ex-ante approach, i.e. only papers that measured farmers' attitudes prior to the adoption of a new technology have been taken into account.

Table 11.3 reports on the factors used in the selected articles to explain the adoption behaviour. We also show their Cronbach alphas' coefficients to compare the quality of the measurement scales of the factors across the sixteen studies. All factors show good coefficients of Cronbach alpha, which means a high reliability.

The factors most used in explaining the adoption behaviour are Attitude toward the Behaviour, Perceived Usefulness (PU) and Perceived Ease of Use (PEOU). Other factors, such as Intention to Use, Subjective Norms and Perceptions of Control (internal and external) also appear as relevant in explaining the behaviour. Even though TPB and TAM are the models that have been adopted mainly in these studies, Subjective Norms emerge as relevant in explaining the behaviour in only four studies. This is surprising, given the fact that this factor is considered as one of the most powerful in explaining the behaviour in theoretical models. Nevertheless, future research in the agricultural domain may deepen the study of the social factors (SC) in the technology adoption behaviour.

Table 11.2 Papers on the adoption of technology in agriculture and their main findings

Nr	Paper	Theoretical model	Method	Technologies	Setting	Sample size	Data source	Main findings
1.	Lynne et al. (1995)	Theory of Planned Behaviour and Theory of Derived Demand	Tobit Regression Model	Water saving technology	Crop growers (strawberry)	44	Florida, USA	Both Perceived and Actual External Control are important for explaining the Behaviour. Policy makers should apply a moderate control on farmer's decisions combining moral suasion and incentives in their policies.
2.	Herath (2013)	Theory of Planned Behaviour	Regression Model	New farm technology	Farmers	36	Southern Moravian region, Czech Republic	Intention, Attitude, Perceived Behavioural Control, Age and Education affects positively the Adoption of technology systems from part of farmers.
3.	Flett et al. (2004)	Technology Acceptance Model	Factor analysis and discriminant function analysis	Dairy farming technologies	Dairy Farms	985	North and South islands, New Zeeland	Farmers consider separately Perceived Ease of Use Perceived Usefulness.
								When deciding about the adoption of a new technology, Perceived Usefulness is more important than Perceived Ease of Use.
4.	Adrian et al. (2005)	Technology Adoption Model	Structural Equation Model	Seven PA technologies	Row crop Growers	85	Alabama, USA	Attitudes of Confidence toward using PA technologies, Perceptions of Net Benefit, Farm Size and Farmer Educational Levels positively influence the Intention to Adopt PA technologies. The Perception of Usefulness positively influences the Perception of Net Benefit.

5.	Rehman et al.(2007)	Theory of Reasoned Action	Principal Component Analysis	Oestrus detection, nitrogen supply management, and, inclusion of white clover	Dairy Farms	145	Cornwell, Devon and Dorset, England	Cost Effectiveness, Improved Detection and Conception Rates are the main drivers when adopting new technologies. Threat of demeaning the personal knowledge and skills of a farmer in 'knowing' their cows is a barrier.
6.	Folorunso and Ogunseye (2008)	Technology Acceptance Model and Model of PC Utilization	Structural Equation Model	Knowledge management system	Farmers, extension specialist and researchers	370	Abeokuta, Ogun state, Nigeria	All original TAM constructs affect computer usage. Social Factors and Facilitating Conditions (constructs from MPCU) are good predictors in the TAM.
7.	Pouratashi and Rezvanfar (2009)	Technology Acceptance Model	Descriptive and inferential analysis	Basic ICT: word processing, spreadsheets, internet access, etc.	Students	110	Teheran, Iran	Skills have direct and indirect effects on the application of Information Communication Technology (ICT). Support and facilities affect the application of ICT indirectly. When students' skills improve, they are more likely to use ICT.
8.	Zhang et al. (2009)	Technology Acceptance Model	Correlational Analysis	Information technology	Farmers	231	Different regions, China	Perceived Usefulness, Perceived Ease of Use, Learning Intention, Risk Preference, and Experience are considered important factors which positively affect farmers' Behaviour in adopting Market Information Systems. Income and Education may also affect the decision.

(continued)

Table 11.2 (continued)

Nr	Paper	Theoretical model	Method	Technologies	Setting	Sample size	Data source	Main findings
9.	Moghaddam and Salehi (2010)	Technology Adoption Model	Structural Equation Model	Yield monitoring, grid soil sampling and the variable rate technologies irrigation, fertilizer, tillage, spraying and seeding	Agricultural Specialists	249	Fars and Khuze-stan, Iran	Trialability has a significant effect on Perceived Ease of Use, Attitude and Intention.
								Observability has direct effect on Perceived Ease of Use, Perceived Usefulness, Attitude and Intention to Use.
								Among all the independent variables, Attitude to Use is the variable that affect the adoption of agriculture technologies mostly.
								Producers who declared confidence about using and learning PA technologies have greater propensity to adopt them.
10.	Yueh and Liu (2010)	Technology Adoption Model	Linear and Multiple Regression Analysis	Farm management information system (FMIS)	Farmers	23	Taiwan	Farmers' skills in Farm Management Information System (FMIS) improved after the training.
								In the future, perceived usefulness could positively affect farmers' motivation to use FMIS.
11.	Aubert et al. (2012)	Technology Adoption Model	Partial Least Square (PLS)	Six PA technologies	Crop Growers (cereal and oleaginous)	438	Quebec, Canada	PA technology adoption is determined by: Perceived Ease of Use and Usefulness, Resource Availability, Trialability, and Voluntariness (negatively), as well as personal characteristics of the farmer (Innovativeness, and Level of Education).
								Farmer's Age and Farm Size do not influence the Adoption.

12.	Sharifzadeh et al. (2012)	Theory of Planned Behaviour	Structural Equation Model	Climate information technology	Crop Growers (wheat)	314	Fars, Iran	Greater Attitude (Instrumental and Affective) toward Use of information in farming decisions was associated with stronger Intention to engage in behaviour. The Modified Theory of Planned Behaviour provides a significant improvement on the model fit by adding a direct causal path from Attitude to Behaviour.
13.	Tey et al. (2014)	Theory of Interpersonal Behaviour and the Theory of Diffusion of Innovation	Logistic Regression Model	Sustainable agricultural practices	Crop Growers (vegetables)	1168	Five regions of Malaysia	Adoption depends on a range of socio-economic, agro-ecological, institutional, informational, and psychological factors, as well as the perceived attributes of sustainable agricultural practices.
14.	Niles et al. (2016)	Theory of Planned Behaviour	Multiple Regression Model	Climate change practices	Farmers	490	Marlborough and Hawke's Bay, New Zealand	Attitudes towards climate change are not significantly associated with change of behaviour. Persisting in stating facts about climate change in hopes to change people's behaviours is not necessarily impactful.
15.	Lu et al. (2015)	Technology Acceptance Model	Partial Least Squares regression	Government-sponsored agricultural information systems	Farmers	1504	Jiangxi, China	Government's role is relevant as it positively affect intention to use new technologies.
								Perceived Enjoyment and Perceived Usefulness are considered important elements for male farmers. Female farmers consider as important only Perceived Usefulness.

(continued)

Table 11.2 (continued)

Nr	Paper	Theoretical model	Method	Technologies	Setting	Sample size	Data source	Main findings
16.	Alavion et al. (2016)	Theory of Planned Behaviour	Multiple Regression Model	E-marketing of agricultural commodities	Agricultural professionals (public and private sectors)	146	Guilan, Iran	Public and private sector professionals consider e-marketing as an important tool for farmers. The three factors of the Theory of Planned Behaviour are confirmed as predictors of the intention to adopt e-marketing. Subjective Norms and Perceived Behavioural Control show higher impact on public professionals, than personal Attitude.

Table 11.3 The constructs measured in each paper and their coefficients of Cronbach Alpha*

Factors	Papers															
	Lynne et al. (1995)	Herath (2013)	Flett et al. (2004)	Adrian et al. (2005)	Rehman et al. (2007)	Folorunso and Ogunseye (2008)	Zhang et al. (2009)	Pouratashi and Rezvanfar (2009)	Yueh and Liu (2010)	Moghaddam and Salehi (2010)	Aubert et al. (2012)	Sharifzadeh et al. (2012)	Tey et al. (2014)	Niles et al. (2016)	Lu et al. (2015)	Alavion et al. (2016)
Adopt/Not adopt (DND)	✓					.82										
Affect Towards Use (ATU)														.65		
Attitude toward Behaviour (AB)	✓	.71			.82	.74		.86		.83		.75	.96		.93	>.70
Biophysical Concern														.87		
Coercive Power (COE)															.95	
Compatibility (COB)											.86		.86			
Complexity (CPX)													.89			
Computer Anxiety (CANX)												.75				
Computer Self-Efficacy (CSE)				.87						.57						
Contact Scale														.65		

(continued)

Table 11.3 (continued)

Factors	Papers															
	Lynne et al. (1995)	Herath (2013)	Flett et al. (2004)	Adrian et al. (2005)	Rehman et al. (2007)	Folorunso and Ogunseye (2008)	Zhang et al. (2009)	Pouratashi and Rezvanfar (2009)	Yueh and Liu (2010)	Moghaddam and Salehi (2010)	Aubert et al. (2012)	Sharifzadeh et al. (2012)	Tey et al. (2014)	Niles et al. (2016)	Lu et al. (2015)	Alavion et al. (2016)
Control Beliefs of New Technology Adoption		.63														
Education Level				✓			✓									
Environmental Policy														.69		
Experience							✓									
Expert Power (EXP)															.89	
Extrinsic Motivation (EM)																
Facilitating Conditions (FC)						.63					.81					
Farm Size				✓												
Income							✓									
Information											.80					
Innovativeness											.83					
Intention to use					✓	.69				.88		.96			.91	
Knowledge (employee)											.84					
Knowledge (farmers)											.72					

Learning Intention							✓									
Legitimate Power (LEG)															.89	
Motivation to comply												.64				
Motivation to comply with others		.84														
Normative beliefs												.74				
Outcome Expectations Personal (OEP)												.75				
Output Quality (OUT)														.65		
Perceived Ease of Use (PEOU)			.64	.88		.83	✓	.86	.92	.89	.84	.70			.92	
Perceived Enjoyment (ENJ)															.90	
Perceived Usefulness (PU)			.82	.93		.77	✓	.86	.92	.58	.89	.82			.92	

(continued)

Table 11.3 (continued)

Factors	Papers															
	Lynne et al. (1995)	Herath (2013)	Flett et al. (2004)	Adrian et al. (2005)	Rehman et al. (2007)	Folorunso and Ogunseye (2008)	Zhang et al. (2009)	Pouratashi and Rezvanfar (2009)	Yueh and Liu (2010)	Moghaddam and Salehi (2010)	Aubert et al. (2012)	Sharifzadeh et al. (2012)	Tey et al. (2014)	Niles et al. (2016)	Lu et al. (2015)	Alavion et al. (2016)
Perceptions of External control (PEC)	✓	.73										.65				>.70
Perceptions of Internal Control (PIC)	✓	.85										.65				>.70
Potential adaptation adoption														.78		
Potential mitigation adoption														.76		
Quality of support											.79					
Referent Power (REF)															.88	
Relative Advantage (RA)				.86							.81		.87			
Result Demonstrability (RES)										.87	.56					
Reward Power (REW)															.88	

Risk preference							✓									
Social Factors (SC)						.57								.75		
Subjective Norm (SN)	✓	.85			✓											>.70
Trialability										.87	.86		.85			
Use (USE)						.82										
Visibility (VIB)										.87						
Voluntariness (VOL)											.60					

The '✓' Indicates the factor has been measured in the paper but, given the particular methodology applied in the research, the Cronbach alpha was not reported
* Cronbach alphas ≥0.70 are considered acceptable (Hair et al. 2010, p. 124)

11.4 Discussions and Conclusion

In general, in the application to agricultural practices, all the reviewed theoretical models find good support. For instance, Folorunso and Ogunseye (2008), in their study of users' acceptance of AGROWIT, a knowledge management information system, found that all the constructs of the TAM were good predictors of the behaviour of participants. This might not be surprising given that TAM was initially developed using data on the evaluation of office technologies. In addition to factors taken from TAM, Folorunso and Ogunseye (2008) included Social factors (SC) and Facilitating Conditions (FC) in their model from the Model of PC Utilization (MPCU) and found a positive effect on the Intention to Adopt PA technologies. In line with this, Adrian et al. (2005) and Moghaddam and Salehi (2010) noticed that Farm Size, Perceptions of Net Benefit, Gender and Technology Awareness were good predictors of Technology Adoption. Aubert et al. (2012) reported that Age and Farm Size did not have any effect on Adoption, whereas other factors such as Perceived Ease of Use, Perceived Usefulness and Resource Availability are good predictors of the Intention to engage in PA.

Regarding the Theory of Planned Behaviour, almost all the theoretical constructs are supported by the studies included in this review. According to Lynne et al. (1995) and Herath (2013), the Intention to Adopt technology practices is related to the Attitude towards the Behaviour, Subjective Norms, Perceptions on Control and other sociodemographic characteristics such as Age and Gender. Despite the fact that Education is not included in the original theory, Herath (2013) revealed that it is a good predictor of Behaviour. Moreover, Sharifzadeh et al. (2012) showed that, among wheat growers, the ones with a positive attitude to use information in their farming decision were more predisposed to implement new and innovative technologies in their farming activity. Finally, in their study about climate change issues, Niles et al. (2016) showed that even though farmers were continuously exposed to possible threats of climate change, it did not have any effect on their behaviour.

It is important to understand the reasons why research focuses on these models and how other models might be applied especially to the field of agriculture. Our review shows that among the theoretical models mentioned in the previous paragraph, TPB and TAM have a greater capability for explaining behaviour. Some final remarks are noteworthy. Despite the variety of factors that the different models on technology adoption consider, all of them share the view that the decision to engage in technology adoption is largely affected by individual perceptions (e.g. Perceived Ease of Use, Attitudes and Perceived Usefulness). When comparing factors in the theoretical models, Experience was found as relevant in explaining the behaviour in six models. Nevertheless, in the literature focused on the adoption of farm technology a central role of experience was not found because studies focused mainly on non-users of technology.

Aubert et al. (2012) found that Voluntariness negatively affects the Intention to Use PA technologies. This suggests that adopters are more influenced by external pressures, considering the use of PA technologies as a legal requirement or a recommendation of their cooperative. Therefore, Aubert et al. (2012) suggest that PA technology adoption could be enhanced by reducing the level of voluntariness. For instance, the introduction of specific regulations (e.g. norms on the use of pesticides and fertilizers) would raise the issue of compulsory reporting required of farmers. Since PA technologies can provide more accurate information for reporting, new regulations could negatively affect voluntariness and increase PA adoption. This conclusion suggests that Voluntariness can be a relevant factor for the adoption of models, but it has not emerged prominently from studies investigating the agricultural domain yet. While Aubert et al. (2012) used voluntariness to explain adoption, their findings might be affected by a reduced level of reliability (Cronbach alpha: 0.60) and the reduction of two items in a factorial scale of five. Other studies might take into consideration voluntariness investigating differences between users and non-users of PA technologies. The example on Voluntariness is particularly interesting because it suggests that other individual attitudes such as Visibility and Motivation to Comply with Others could also have a similar effect on technology adoption. These factors have scarcely been included in the models on PA technology adoption so far, and they could be explored as possible mediators between contextual factors and the intention to adopt or the perceptions on usefulness or on the ease of use. For instance, Aubert et al. (2012) suggested that the relation between environmental policy and adoption was mediated by Voluntariness. Similarly, other individual attitudes could play the role of mediators between the availability and quality of support, environmental policy or other facilitating conditions and PA adoption or to the Perceived Ease of Use or the Perceived Usefulness of technology.

Since TAM has been used mainly to study the intention to adopt information technologies, its application in the evaluation of technologies with a significant orientation on information processing seems worthy. Nevertheless, the Model of PC Utilization and the Unified Theory of Acceptance and Use of Technology may go beyond the individual level considered in the TAM to take into account social and environmental features as facilitators in the approach to technology adoption. In fact, what remains surprisingly overlooked is the effect that the others may exert on individuals' stance towards technology. Studies on social networks and innovation have long underlined the role that interactions with peers may play in the adoption and diffusion of technology (e.g. Burt 1980; Tucker 2008). The opinions and experience concerning the technology of those with whom one interacts are able to affect the orientation in technology use. We can follow the example of individuals who are similar to us in terms of relations or, alternatively, we can imitate peers to whom we are strongly linked.

Studies addressing technology adoption in agriculture do not so far allow us to understand how individuals who have different access to network resources may react when appraising technology adoption. Independent farmers might act differently from farmers operating within an integrated value chain and supply network. Similarly, farmers who cooperate with large distributors may select courses of action, when gauging technological opportunities, which differ from those available to farmers cooperating with micro-companies to serve local markets. An integration of the models with a social network perspective could enrich the understanding of the processes through which adoption is evaluated, chosen and implemented in an industry characterized by heterogeneous patterns of social ties.

In addition, in the development of a scale for the evaluation of entrepreneurship in agriculture and food production, in addition to the financial, human-related and technological factors, it was found that infrastructure and network elements are constituents of the measurement instrument (Bolzani et al. 2016). The infrastructure factor refers to the availability of tangible and intangible resources for knowledge development and sharing. They can range from R&D facilities to mentoring and counselling organizations (e.g. Knudson et al. 2005). The presence of an infrastructural layer in the innovation system might favour the overcoming of some factors that hinder the diffusion of innovative PA technologies. Agricultural innovation is often capital intensive and requires a good level of training on the farmer's side. In the European context, made of smallholders, the existence of social ties between farmers could improve their bargaining power and their influence on the development of infrastructural entities. These latter may be able to create the necessary economies of scale and scope to close the knowledge gap faced by farmers. Specialized support could in fact become accessible where a consistent level of demand sustains its development.

Finally, our review focused on ex-ante approaches to appraise how an innovation can be perceived in the agricultural domain and raise the interest of farmers. As a future research direction, we shall devise an investigation on how different public policies could foster the introduction of innovation in agriculture. In many countries, local, national and transnational institutions enact initiatives to support the introduction of PA. These policies can be evaluated in terms of their efficacy in favouring the diffusion of the innovations and efficiency in the use of public funds. We suggest that both in their design and assessment, the complex process of adoption that we described in this chapter should be taken into consideration.

References

Adrian AM, Shannon HN, Paul LM (2005) Producers' perceptions and attitudes toward precision agriculture technologies. Comput Electron Agric 48(3):256–271

Ajzen I (1991) The theory of planned behaviour. Organ Behav Hum Decis Process 50:179–221

Ajzen I, Fishbein M (1980) Understanding attitudes and predicting social behavior. Prentice-Hall, Englewood Cliffs

Alavion SJ, Allahyari MS, Al-Rimawi AS, Surujlal J (2016) Adoption of agricultural E-marketing: application of the theory of planned behavior. J Int Food Agribusiness Market. doi: org/10.1080/08974438.2016.1229242

Aubert BA, Andreas S, Jonathan G (2012) IT as enabler of sustainable farming: an empirical analysis of farmers' adoption decision of precision agriculture technology. Decis Support Syst 54(1):510–520

Bandura A (1986) Social foundations of thought and action: a social cognitive theory. Prentice-Hall, Englewood Cliffs

Bolzani D, Carli G, Fini R, Sobrero M (2016) Promoting entrepreneurship in the agri-food industry: Policy insights from a pan-European public–private consortium. Ind Innov 22(8):753–784
Burt RS (1980) Models of network structure. Annu Rev Sociol 6(1):79–141
Census of Agriculture (2012) Preliminary report highlights. Farms and farmers. https://www.agcensus.usda.gov/Publications/2012/Preliminary_Report/Highlights.pdf
Compeau DR, Higgins CA (1995) Computer self-efficacy: development of a measure and initial test. MIS Q 19:189–211
Davis FD (1989) Perceived usefulness, perceived ease of use, and user acceptance of information technology. MIS Q 13(3):319–340
Davis FD, Bagozzi RP, Warshaw PR (1989) User acceptance of computer technology: a comparison of two theoretical models. Manag Sci 35(8):982–1003
Davis FD, Bagozzi RP, Warshaw PR (1992) Extrinsic and intrinsic motivation to use computers in the workplace. J Appl Soc Psychol 22(14):11–32
Deci EL (1971) Effects of externally mediated rewards on intrinsic motivation. J Pers Soc Psychol 18:105–115
Deci EL (1975) Intrinsic motivation. Plenum, New York
Deci EL, Ryan RM (1985) Intrinsic motivation and self-determination in human behavior. Plenum, New York
Farm structure statistics – Statistics Explained [WWW Document] (2013) URL http://ec.europa.eu/eurostat/statistics-explained/index.php/Farm_structure_statistics#The_size_of_agricultural_holdings. Accessed 12 Dec 2016)
Fishbein M (1967) A behaviour theory approach to the relations between beliefs about an object and the attitude toward the object. In: Fishbein M (ed) Readings in attitude theory and measurement. Wiley, New York, pp 389–400
Fishbein M, Ajzen I (1975) Belief attitude, intention and behavior: an introduction to theory and research. Addison-Wesley, Reading
Flett R, Alpass F, Humphries S, Massey C, Morriss S, Long N (2004) The technology acceptance model and use of technology in New Zealand dairy farming. Agric Syst 80:199–211
Folorunso O, Ogunseye SHO (2008) Applying an enhanced technology acceptance model to knowledge management in agricultural extension services. Data Sci J 7:31–45
Hair JF, Black WC, Babin BJ, Anderson RE (2010) Multivariate data analysis, 7th edn. Prentice Hall, Upper Saddle River
Harrison DA, Mykytin PP, Riemenschneider CK (1997) Executive decisions about adoption of information technology in small business: theory and empirical tests. Inf Syst Res 8(2):171–195
Hartwick J, Barki H (1994) Explaining the role of user participation in information system use. Manag Sci 40(4):440–465
Herath CH (2013) Does intention lead to behaviour? a case study of the Czech Republic farmers. Agric Econ 59(3):143–148
Karahanna E, Straub DW, Chervany NL (1999) Information technology adoption across time: a cross-sectional comparison of pre-adoption and post-adoption beliefs. MIS Q 23(2):183–213
Knudson W, Wysocki A, Champagne J, Peterson HC (2005) Entrepreneurship and innovation in the agri-food system. Am J Agric Econ 86(5):1330–1336
Lawler EE, Porter LW (1967) Antecedent attitudes of effective managerial performance. Organ Behav Hum Perform 2:122–142
Lu Y, Lu Y, Wang B, Pan Z, Qin H (2015) Acceptance of government-sponsored agricultural information systems in China: the role of government social power. IseB 13:329–354
Lynne GD, Casey FC, Hodges A, Rahmani M (1995) Conservation technology adoption decisions and the theory of planned behavior. J Econ Psychol 16(4):581–598
Mathieson K (1991) Predicting user intentions: comparing the technology acceptance model with the theory of planned behavior. Inf Syst Res 2(3):173–191
Moghaddam RK, Salehi S (2010) Agricultural specialists' intention toward precision agriculture technologies: integrating innovation characteristics to technology acceptance model. Afr J Agric Res 5(11):1191–1199

Moore GC, Benbasat I (1991) Development of an instrument to measure the perceptions of adopting an information technology innovation. Inf Syst Res 2(3):192–222
Morris MG, Venkatesh V (2000) Age differences in technology adoption decisions: Implications for a changing work force. Pers Psychol 53(2):375–403
Niles MT, Brown M, Dynes R (2016) Farmer's intended and actual adoption of climate change mitigation and adaptation strategies. Clim Chang 135(2):277–295
Pierpaoli E, Carli G, Pignatti E, Canavari M (2013) Drivers of precision agriculture technologies adoption: a literature review. Procedia Technol 8:61–69
Pignatti E, Carli G, Canavari M (2015) What really matters? a qualitative analysis on the adoption of innovations in agriculture. J Agric Inform 6(4):73–84
Pouratashi M, Rezvanfar A (2009) Analysis of factors influencing application of ICT by agricultural graduate students. J Am Soc Inf Sci Technol 61(1):81–87
Rehman T, Kemey MK, Yates CM et al (2007) Identifying and understanding factors influencing the uptake of new technologies on dairy farms in SW England using the theory of reasoned action. Agric Syst 94(2):281–293
Rogers EM (1995) Macmillian Publishing Co. Diffusion of innovations. http://hollis.harvard.edu/?itemid=%7Clibrary/m/aleph%7C006256656
Rogers EM (2003) Macmillan Publishing Co. Diffusion of innovations. https://teddykw2.files.wordpress.com/2012/07/everett-m-rogers-diffusion-of-innovations.pdf
Rogers EM, Shoemaker FF (1971) Communication of innovations: a cross-cultural approach. Free Press, New York
Schimmelpfennig D (2016) Farm profits and adoption of precision agriculture, ERR-217, U.S. Department of Agriculture. Economic Research Service
Sharifzadeh M, Zamani GH, Khalili D (2012) Agricultural climate information use: an application of the planned behaviour theory. J Agric Sci Technol 14:479–492
Sheppard BH, Hartwick J, Warshaw PR (1988) The theory of reasoned action: a meta-analysis of past research with recommendations for modifications and future research. J Consum Res 15(3):325–343
Taylor SH, Todd P (1995a) Assessing IT usage: the role of prior experience the influence of prior experience. Manag Inf Syst 19:561–570
Taylor SH, Todd P (1995b) Understanding information technology usage: a test of competing models. Information systems research
Tey YS, Li E, Bruwer J (2014) The relative importance of factors influencing the adoption of sustainable agricultural practices: a factor approach for Malaysian vegetable farmers. Sustain Sci 9(1):17–29
Thompson RL, Higgins CA, Howell JM (1991) Personal computing: toward a conceptual model of utilization. MIS Q 15(1):124–143
Tornatzky LG, Klein JK (1982) Innovation characteristics and innovation adoption-implementation: a meta-analysis of findings. IEEE Trans Eng Manag 29(1):28–43
Triandis HC (1980) Values, attitudes and interpersonal behavior. In: Howe HE (ed) Nebraska symposium on motivation, 1979: beliefs, attitudes and values. University of Nebraska Press, Lincoln, pp 195–259
Tucker C (2008) Identifying formal and informal influence in technology adoption with network externalities. Manag Sci 54(12):2024–2038
Vallerand RJ (1997) Toward a hierarchical model of intrinsic and extrinsic motivation. Adv Exp Soc Psychol 29(27):271–360
Venkatesh V, Bala H (2008) Technology acceptance model 3 and a research agenda on interventions. Decis Sci 39(2):273–315
Venkatesh V, Davis FD (2000) A theoretical extension of the Technology Acceptance Model: four longitudinal field studies. Manag Sci 46(2):186–204
Venkatesh V, Morris MG (2000) Why don't men ever stop to ask for directions? gender, social influence, and their role in technology acceptance and usage behaviour. MIS Q 24(1):115–139

Venkatesh V, Speier C (1999) Computer technology training in the workplace: a longitudinal investigation of the effect of mood. Organ Behav Hum Decis Process 79(1):1–28
Venkatesh V, Davis FD, Smith RH, Walton SM (2000) A theoretical extension of the technology acceptance model: four longitudinal field. Manag Sci 46(2):186–204
Venkatesh V, Morris MG, Davis GB, Davis FD (2003) User acceptance of information technology: toward a unified view. MIS Q 27(3):425–478
Vroom V (1964) Work and motivation. Wiley, New York
Yueh HP, Liu YL (2010) Effects of farmer's computer abilities and self-efficacy on their learning performance and adoption intention of the farming management information system. J Agric Ext Rural Dev 2(9):191–197
Zarco-Tejada PJ, Hubbard N, Loudjani P (2014) Precision agriculture: an opportunity for Eu farmers – potential support with the cap 2014–2020. Directorate-general for Internal Policies Policy Department B: Structural and Cohesion Policies Agriculture and Rural Development. http://www.europarl.europa.eu/RegData/etudes/note/join/2014/529049/IPOL-AGRI_NT(2014)529049_EN.pdf
Zhang J, Zhang X, Mu W, Zhang J, Fu Z (2009) Farmers' information usage intention in China based on the technology acceptance model. Comput Comput Technol Agric II 3:1845–1853

Chapter 12
Perspectives of Precision Agriculture in a Broader Policy Context

Kim Martin Lind and Søren Marcus Pedersen

Abstract Agriculture is faced with contrasting requirements from the broader society. On the one hand, agriculture needs to expand production to be able to feed a growing global population. Furthermore, the developing bio-economy requires agriculture to produce for a range of non-food objectives such as bio-fuel, textile fibres, etc. On the other hand, concerns over the environment, climate, biodiversity and other ecosystem services place restrictions on conventional agricultural production. Precision agriculture can be part of the response to these often conflicting issues by employing technologies that in a precise and targeted approach reduce resource use and increase yield. Furthermore, the growing demand for higher value food products in terms of health and quality require traceability and information about production processes and resource use, which also correspond with the possibilities offered by precision agriculture technology. The general movement towards higher integration in food supply chains is a natural extension of the requirements for traceability and product information, which are integral parts of precision agriculture.

Keywords Precision agriculture • Public goods • Societal trends • Adoption

12.1 Introduction

Precision agriculture (PA) promises to provide a better and more targeted use of inputs and reduce the negative effects to the environment. In particular, auto-steering and the availability of Variable-Rate Application (VRA) equipment at affordable costs produces some promising perspectives. With growing prices of agrochemicals, VRA provides economic benefits to farmers while providing ecological

K.M. Lind (✉) • S.M. Pedersen
Department of Food and Resource Economics, Faculty of Science, University of Copenhagen, København, Denmark
e-mail: kim@ifro.ku.dk; marcus@ifro.ku.dk

S.M. Pedersen, K.M. Lind (eds.), *Precision Agriculture: Technology and Economic Perspectives*, Progress in Precision Agriculture,
https://doi.org/10.1007/978-3-319-68715-5_12

benefits with reduced leaching and emissions. In this chapter, we envisage how PA may attain larger yields with lower inputs to meet the standards and policy trends in Europe and other OECD countries.

In the next 20 years, world food production is required to increase by 50% to feed the projected 9.2 billion Earth population in 2050, (FAO 2009). Up to 80% of that increase must come from production intensification. Precision agriculture can be one of the options to deal with the growing demand for food. The size of this variability can be used to demonstrate the suitability of implementing VRA more extensively. Although site-specific techniques have been available to farmers for more than 20 years, the adoption has been relatively slow for most applications. So far, variable-rate N application in cereals is used by 3–8% only of the farmers in Denmark, Germany and Finland (Lawson et al. 2011).

12.2 PA and Wider Societal Trends

Precision Agriculture technologies offer promising perspectives on meeting the demands of and mitigating risks to the global society. In particular, the growing world population requires increases in food production, however, concerns over climate and environment may slow down the growth necessary in productivity. Furthermore, there are concerns about increasing scarcity of and decreasing returns to conventional energy reserves, which in turn could significantly reduce the amount of affordable energy for societal needs and demands. Climate change poses threats to sustainability of natural environments, agriculture and urban areas. Agricultural production for non-food uses, specifically for energy production, requires additional agricultural area in addition to areas needed for expanding food production, for habitation and other urban related activities for a growing population that is increasingly living in large cities.

Rabbinge and Bindraban (2012) identify **six megatrends** in agriculture that overarches global development albeit with varying local or regional effects. The **increase in productivity** is identified as the **first megatrend**. Hitherto, the increases in food production have outpaced the increase in population despite the dire predictions of Malthus (1798). Thus, over decades if not centuries productivity in agriculture has increased in terms of per hectare, per hour of labour, per kg of chemicals applied and for other inputs. In the Netherlands, wheat yields have risen from around 800 kg per hectare in 1400 to 1800 kg in 1900 and increasing to 9000 kg per hectare in 2000, (Bindraban and Rabbinge 2012). Simultaneously, labour input has decreased from about 600 h per hectare in 1400 to 240 h in 1900 and dropping to 12 h per hectare in 2000. Behind these impressive achievements are better and more targeted uses and applications of inputs. Information gathering of the conditions of crops in the fields has led to the ability to vary input and resource use according to differing needs in line with the principles of precision agriculture.

Farmers are generally price takers – meaning that it is difficult for the individual farmer to get a price different from what the market offers. Adoption of a new farming technique enables the individual farmer to reduce cost and or improve yields and thereby profits in the short run. However, after some time other farmers may adopt the techniques and the aggregate output of a given product will increase. Without an increase in demand, prices will fall. In reaction to this, new technologies will be developed to reduce costs. These technologies will be commercialized and adopted by farmers, which in turn provide new supplies and a reduction in output prices. This course of events is known as the agricultural treadmill. It is often observed in the agribusiness sector because most agricultural products are regarded as primary products produced by many producers. As an individual farmer, it can profitable to be among the early adopters of new technology enabling possible favourable returns in the short run (Doll and Orazem 1978). However, in the longer run above-normal profits will decline as more and more farmers implement the technology.

Innovations are often caused by changes in relative factor prices. With higher labour costs, the agricultural sector will be forced to use more capital-intensive factors as a substitute for labour to gain an increase in productivity and profitability. The relative cost of capital compared with labour has shifted the agricultural sector into more specialization with larger farm areas and production units. This trend has been prevalent in Europe and many other regions since the fifties (Pedersen 2003).

Precision agriculture and smart farming technologies have to some extent followed a similar pattern. Precision farming is capital rather than labour intensive and the concept of auto steering and variable treatment aims at saving variable inputs such as fuel, nutrients and to some extent labour and thereby increasing farm productivity.

The **second identified megatrend** in agriculture is the **integration of more advanced industrial and information technologies** in agricultural production. Increasingly, farm machinery is equipped with sensors and GNSS capabilities making information gathering and processing a more and more natural element of farming practices. This development has contributed towards making it possible to identify spatial and temporal variability across fields, soils, crops, pests and weed infestations, and management practices. Precision agriculture technology is considered by farmers mainly because of higher expected profitability, (Reichardt and Jürgens 2008). The main reason for the low rate of adoption of precision agriculture in Germany was found to be the high cost of the technology. Nevertheless, more and more PA-technologies such as positioning systems and sensors are embedded in new farm machinery and equipment. Therefore, gradually over time it is expected that PA-technologies will increasingly be adopted with ongoing investments. Figure 12.1 shows the evolution of the number of tractors in Europe, USA, Brazil, China and Sub-Saharan Africa (SSA), where tractor density on agricultural land is used as a proxy for the intensity of capital investments in agriculture, (Jepsen et al. 2015).

The figure shows that investments in technology can take place fairly rapidly. In Europe, the number of tractors has increased from 3 to 8 per 100 ha of arable land from 1960 to 1990 as a part of the mechanization in Europe. In 2005, this number has been reduced to 7 tractors per 100 ha in Europe. Within Europe there are major differences because of differences in the farm area per farmer amongst different European countries. With relatively large farms in for instance the UK, France and some East European regions and small farms in Southern Europe.

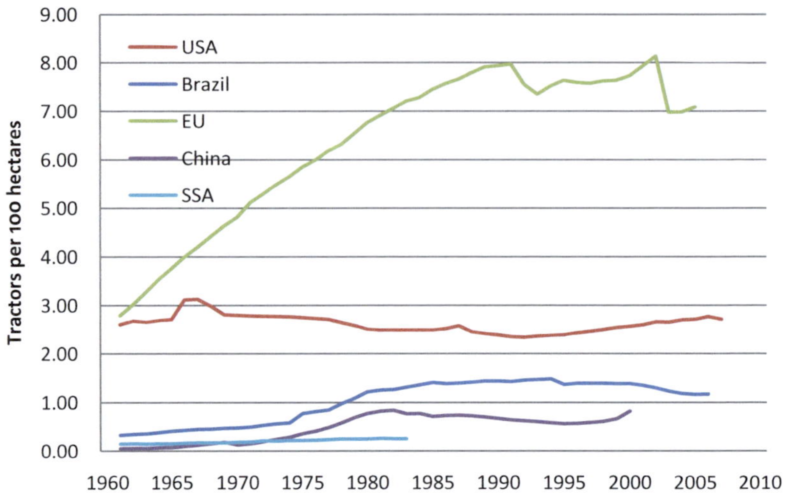

Fig. 12.1 Tractors per 100 ha of arable land (Source: World Bank Development Indicators)

In South America, Brazil more than doubled the number of tractors per hectare in the 1970s. Likewise, the USA can also be seen to have substantial technological investments in certain periods, but on average the number of tractors have been between 2–3 tractors per 100 ha, which to some extent is caused by relatively large farm holdings in the USA. Hence, when the technology is considered beneficial or profitable enough, farmers are willing to undertake significant investments.

This development is not only about increasing numbers of larger and larger farm vehicles. The current development of computers and microchips is likely to speed up the development in precision agriculture. Moore's law says that the number of transistors in a dense integrated circuit doubles approximately every 2 years. Basically, it means that the capacity of microprocessors doubles every second year. A similar trend is seen with regard to the number of pixels per unit costs and other developments in computers and electronics. This trend will provide opportunities for the development of smart farming technology that does not rely exclusively on large machines. In addition, big data have become a term for gathering, storing and analysing large and complex data sets, Marz and Warren (2015), which is being used to analyse and interpret enormous data files at high speeds. Eventually, it is expected that various agricultural processes will benefit from progress in data processing techniques such as weed recognition, soil mapping, plant requirements, etc. Furthermore, this trend is likely to lead to more advanced systems in the future such as autonomous systems at progressively lower costs.

The development towards **integration of the whole food supply chain** is identified as a **third megatrend**. This development enables producers, processors and distributors to comply with sanitary and phytosanitary standards, reduce environmental impacts and target consumer requirements thereby increasing value added in the supply chain. *Traceability* has become a tool for securing safety and quality of the food products in addition to adding value for customers, who increasingly demand information of location and process characteristics, (Dabbene et al. 2014).

Precision agriculture enables retailers and final consumers to trace and control each action in the supply chain and on the field. Traceability has become an argument by manufacturers of GPS-related equipment for adopting precision farming technologies. Although, PA can be a tool to trace the commodity from "field to fork", it will still require an effort to follow and certify the commodity vertically in the food supply chain. From other farm commodities, we have experienced a price premium for organic products and crops with certain local brands. Similarly, it may be possible to obtain a premium for certifying traceability. Nevertheless, it is a presumption that the final consumers are willing to pay for certified traceability. From a stakeholder workshop in Denmark it was concluded that consumers may find PA too complex to understand and explain as a concept of value. Participants at the workshop found it difficult to "brand the concept of precision farming" in the supermarkets compared with for instance organic products (Pedersen et al. 2002).

Other ways of obtaining an extra premium could be from **selective harvesting** if there is a timely variation in crop quality and maturity. In that case, PA can help to identify which subfield is ready to be harvested in order to obtain higher prices of the final product. Selective harvesting requires optimized route planning systems and sophisticated models to predict crop harvest time.

Multifunctionality of agriculture forms a **fourth megatrend**. Agriculture produces a number of outputs in addition to the immediate production objective of the farmer. Consequently, the farmer is required to meet environmental and other objectives demanded by society. These objectives include biodiversity, landscape management, animal welfare, rural settlement and other public goods. In developed countries, such concerns are increasingly shaping agricultural policies, see e.g. Rizoy (2004), where traditional agricultural support is reallocated towards provision of public goods and increased sustainability of agriculture. *Sustainability* issues are high on the political agenda. Precision agriculture has a strong potential to help agricultural policy to meet its objectives by enhancing competitiveness and improving sustainability and effectiveness (i.e. reducing agriculture's impact on the environment as well as using natural resources in a sustainable manner) (EP 2014).

Europe provides 25% of cereal production worldwide (FAO 2012) and winter wheat (*Triticum aestivum* L.) is the most important crop produced in Europe on 56 million hectares. In Europe, the total use of nitrogen is about 20 million tons of which the cereal production uses more than 10 million tons. Currently, most nitrogen application in winter wheat is carried out as uniform application, and often by using the most demanding part of the crop to define the rate of N-application, which often causes leaching or emissions on areas where the nitrogen is not taken up (Robertson et al. 2008). To achieve large grain yields, N additions are necessary, especially in areas with small soil organic matter content. A study from the Netherlands has shown that with the implementation of variable-rate application techniques, cereal yield can be increased by 10% (D. van der Schans et al. 2008). Other studies have shown that by changing the application from uniform to site-specific application based on measured crop needs, savings can be realised in the

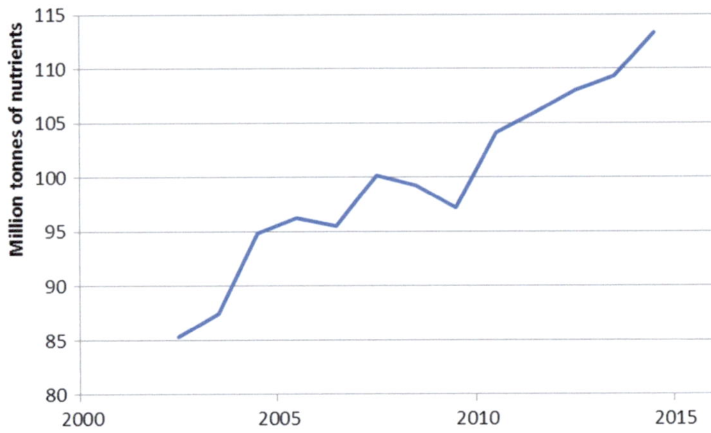

Fig. 12.2 Global production of nitrogen fertilizers 2002–2014 (Source: FAOSTAT)

order of 5% and up to 50% of N in cereals (Scharf et al. 2011; Basso et al. 2012), depending on local soil and production conditions. Additional benefits could come from a reduced usage of nitrogen as well as an increase in grain protein content. Other studies from Denmark have shown little or moderate yields from variable-rate N application based on crop and soil simulation models (Pedersen 2003).

Environmental stress generally increases with more intensive agriculture, which in turn is seen as a prerequisite for productivity increases. Consequently, chemicals such as pesticides, herbicides, insecticides in addition to fertilizers are used increasingly. Figure 12.2 shows the evolution of the global production of nitrogen fertilizers.

Public pressure on the agricultural sector to reduce the negative environmental impact from nitrate leaching, excessive supplies of pesticides and water will increase. Several legal regulations in different regions have been imposed, including quotas and levies on nitrogen application, treatment indices and taxes for pesticides, time limits for irrigation, and recently more focus has been put on phosphorus application (Gachango et al. 2015a; Pedersen 2003 and Pedersen et al. 2013).

The figure shows an increase in nitrogen production of more than 2 million tonnes per year on average. This growth may be necessary to secure the increases in global food production that is required to meet the demands of a growing global population, FAO (2009), however, the substantial yearly increases in nitrogen use presents challenges for the environment and for water resources.

Western nitrogen fertilizers in different forms imported into Europe increased from 6.2 million tons in 2002 to 10.6 million tons in 2014 and Western Europe is currently a net importer of nitrogen (FAOSTAT).

Generally, nitrogen is applied uniformly across fields regardless of site-specific needs and balances. Precision agriculture offers a more targeted approach where sensors detect and identify nitrogen deficiencies much more precisely, which potentially leads to reductions in redundant nitrogen use without complementary decreases in yield. Hence, PA can alleviate stress to the environment and help secure more sustainable agricultural production.

Another example is water consumption in agriculture, (Fraiture and Wichelns 2010). The global demand for water in agriculture will increase over time with increasing population, rising incomes, and changes in dietary preferences. Moreover, increasing demands for water by industrial and urban users, and water for the environment will intensify competition of the limited resources. Precision irrigation can reduce water consumption in high-value crops significantly. In a study of water usage in a citrus orchard (González-Dugo et al. 2013), precision irrigation was able to reduce water use by 25% without reductions in yield. Especially in dry and semi-dry areas, site-specific irrigation is likely to get more attention.

The **fifth megatrend** in agriculture identified by (Rabbinge and Bindraban 2012) concerns **food and health issues**. More and more food consumption is linked to health issues, (Szakaly et al. 2011). Food diets are designed towards specific uses based on diseases, deficiencies and genetic traits. Management of crops and animal husbandry can be improved and optimized by using information gathering sensors mounted on agricultural machinery. Individual animals and plants can, thereby, be monitored and conditional actions related to plant and animal health can be automated or form part of the basis for subsequent farm manager decision making, (OECD 2016). This development is projected to evolve into the management of product quality with increased value added in agricultural products. As precision agriculture may help to reduce nitrate leaching into ground waters, it can be a means to improve the quality of table and drinking water. It may further enable farmers to reduce the application of pesticides, which in turn also affects crop quality and improves drinking water.

The increasing attention given to sustainable agriculture is in part a response to what is seen as the harmful effects of large-scale industrial agricultural systems on the environment and on human health, Horrigan et al. (2002); Gold (2016). Sustainable agriculture includes organic practices and focuses on relatively small integrated farms with less reliance on chemical inputs. Precision Agriculture provides opportunities for decreasing chemical inputs through e.g. site-specific applications and mechanical weeding. Nevertheless, industrial agriculture has achieved high yields, which are reduced in sustainable and organic farming practices. Thus, more land is needed to provide similar production quantities. This is a crucial point in the debate concerning whether land for nature and biodiversity should be segregated from land for production or whether these considerations should be integrated into the production systems, Tscharntke et al. (2012).

The **bio-based economy** is a **sixth megatrend**. A public goods-oriented bio-based economy is based on production paradigms that rely on biological processes and, as with natural ecosystems, use natural inputs, expend minimum amounts of energy and do not produce waste because all materials discarded by one process are inputs for another process and are subsequently reused in the ecosystem, EC (2011). A prominent example is the expansion of bio-based energy production such as ethanol production however, biological products have a variety of different uses which are being explored, Vanholme et al. (2013).

Precision agriculture fits naturally into several of these megatrends. Increasing productivity and reducing resource use is at the root of precision farming. Identifying spatial and temporal variation provides possibilities for the targeted use of resources instead of uniform application, which can lead to significant savings, and is an integral part of the procedures for identifying potential resource reductions. Furthermore, precision agriculture can provide much of the information gathering and traceability necessary for improved integration in the food supply chain. Consequently, the first three megatrends identified in agriculture are well in line with the objectives and capabilities of precision farming technology. Precision agriculture has a strong potential to function as a vehicle allowing farmers to achieve the objectives set by policy makers and society in general concerning environmental and climate issues while simultaneously improving efficiency and competitiveness.

Precision agriculture has the potential to reduce resource use and increase yields by enabling farmers to collect information and improve farm management through better decision-making processes. In particular, variable-rate application promises to reduce environmental stress by using chemicals, fertilisers, water and other resources in a targeted approach. However, variable-rate application technology has yet to demonstrate significant economic benefits for farmers leading to low investment in this technique. Possible environmental gains are often not priced in the markets, which can justify support to obtain the wider societal benefits of positive externalities and public goods production associated with PA. This would be in line with the changing objectives of the Common Agricultural Policy in the EU over the last couple of decades.

Nitrogen is a key to increase productivity and economic returns in crop production. Nitrogen Use Efficiency (NUE) by crops is globally low (between 30 and 50%) (Baligar et al. 2001; Delgado et al. 2010). The need to improve nitrogen efficiencies aimed at reducing the negative environmental impact from losses of nitrogen have been emphasised in relation to surface water, leaching and atmospheric loss, which again contribute to climate change (Li et al. 2007; Dubrovsky et al. 2010; IPCC 2007).

Eutrophication problems in surface water impose negative effects on ecosystems. In practice, only about half of the N fertiliser that is added to today's cropping systems is taken up by the crop (Smil 1999; Robertson and Vitousek 2009). The other 50% remains in the soil or seeps out through the air or water pathways (Mosier et al. 2001).

Fertilisers are important to intensify agricultural production and to ensure food security for the growing population. However, the general public will no longer accept the negative environmental consequences of using fertilisers.

There is a pressure among farmers to **comply with the legislation** (such as Water Framework Directives, Nitrates Directive and River Management Plans) as well as national legislation.

Images from remote sensing show large differences in canopy development that subsequently lead to variation in yield (Primicerio et al. 2012). Crop yield, protein content and nitrate leaching are all functions of nitrogen application. A goal for the farm manager is to improve the financial viability of the farm. For the surrounding society, the goal is to improve overall welfare by reducing negative environmental effects such as nitrate leaching and nitrous oxide emissions. Nitrate leaching from root zone to water and streams is a common consequence of intensive crop production, and the higher the application amounts the greater the leaching. All EU countries are, according to the EU Water Framework Directive, obliged to set up river management plans for their river catchment areas. These plans are detailed descriptions of how targets that are set for the catchment and river basins are to be reached. A recent analysis from the Danish catchment area "Limfjorden" has shown that the marginal costs of reduced nitrogen nitrate leaching from agriculture is around 20–30 € per kg nitrogen N leached to the recipient (Ørum and Jacobsen 2013). Similar costs were found in (Gachango et al. 2015b); The leaching may be regulated by using nitrogen quotas, set aside or other means or by using Variable Rate Nitrogen Application (VRA).

A potential reduction of 2–4 kg N leaching per ha from variable-rate N application with considerable soil variation may be possible. This might lead to an environmental monetary value of EUR 40–120 per ha (Pedersen and Pedersen 2002). Several studies on the performance of VRA have focused on static or historic differences in crop yield and soil type. However, a few models include risk assessment in relation to future precipitation and vegetation indices measured in real time during the growing season. By including these variables it might be possible to reduce nitrate-N leaching even further with more sophisticated models.

12.3 Policy Trends in Europe

Precision agriculture might be able to contribute to achieving the objectives of the EU common agricultural policy by integrating at a large scale across Europe a group of advanced technologies that will enhance resource use efficiency (N fertilizer), while increasing grain quality and yields. Precision farming could potentially contribute to improved farm incomes in addition to improving the competitiveness of the agricultural sector by enabling agriculture to improve efficiency. Through the integration of existing PA technologies, PA may foster green growth through innovation, highlighting the role of agriculture in preserving natural resources and in contributing to the solution of global environmental challenges as well as preserving local natural habitats and environmental goods.

In Europe, the **CAP-reform** process beginning in 1992 has increasingly changed the focus of agricultural policies from traditional production support towards broader societal goals based on multifunctionality and sustainability, Jensen et al. (2009). The CAP has been restructured to two pillars, where the first covers traditional agricultural support and the second is support founded on community preferences. Thus, pillar 2 policies are implemented through national or regional rural development programmes, which are based upon at least four of the six common EU priorities, EC (2013):

1. Fostering knowledge transfer and innovation in agriculture, forestry and rural areas.
2. Enhancing farm viability and competitiveness of all types of agriculture in all regions and promoting innovative farm technologies and sustainable management of forests.

3. Promoting food chain organisation, including processing and marketing of agricultural products, animal welfare and risk management in agriculture.
4. Restoring, preserving and enhancing ecosystems related to agriculture and forestry.
5. Promoting resource efficiency and supporting the shift towards a low carbon and climate resilient economy in agriculture, food and forestry sectors.
6. Promoting social inclusion, poverty reduction and economic development in rural areas.

The set of priorities listed and the range of measures under pillar 2 listed in **Regulation (EU) No 1305/2013 of the European Parliament and of the Council of 17 December 2013** offer opportunities for supporting investment in and dissemination of precision agriculture technology. Moreover, the regulations open possibilities for supporting knowledge transfer, advisory services, training and extension by which information of PA technology and decision support tools can be disseminated among farmers. The requirements for documentation of environmentally-friendly agricultural production for instance in relation to the EU directives concerning nitrate and pesticides provide arguments for supporting implementation of PA technology. Specifically, variable-rate application technology has hitherto not been shown to provide appreciable economic benefits, which would otherwise induce investments by farmers in this technology. Therefore, if the societal values from the environmental and climatic benefits that can be obtained by variable-rate application are to be realised, subsidies for this technology are needed and could be provided through pillar 2 measures. The economic benefits of controlled traffic farming and auto-guiding systems are documented better, thus these technologies should disseminate organically across agriculture.

Precision agriculture will contribute to **a more resource efficient Europe** because it will increase N efficiency and fuel efficiency, reducing reliance on the import of fertilizers that is very energy-dependent. Consequently, the application of technologies that improve N use efficiency is important to improve raw material supply in Europe.

- **Precision agriculture**: directly addresses major EU policies such as the Water Framework Directive, the Nitrates Directive, the Ground Water Directive and the Common Agricultural Policy. Precision agriculture will in its most advanced form become a decision making system that integrates real-time crop status information, data based on field history, weather and economic forecasts and a web-based agronomic decision support system. Integration of all these variables will result in improved and better decision making capacity for farm managers.
- **Traceability**: Precision agriculture will allow the farmer to trace the amount of N used, facilitating record-keeping and compliance with EU environmental regulations including vulnerable areas.
- **Usability:** Precision agriculture places the farmer at the centre of the system by providing a real-time service specifically tailored to the end-user needs and aimed at facilitating decision making, but still relying on farmers' perception.

- **Module-ability:** Precision agriculture will integrate several pre-existing modules that could be substituted or improved according to the end-users' needs and technology availability (local crop models, local decision making systems, local UAV providers.

The need for a European approach is justified by: nitrogen fertilization has a high priority in the European environmental legislation as reflected in the **Water Framework Directive,** which describes the targeted level of good qualitative status of all water bodies such as a good biological status and a good chemical status, and The **Nitrates Directive** that aims to reduce pollution by nitrate leaching.

Furthermore, the **Common Agricultural Policy** and in particular the **cross compliance scheme, the greening measures** and **the rural development programme** couple subsidies to farmers with mandatory minimum levels of agro-environmental criteria.

According to the EU Directive on the sustainable use of pesticides, it is stated that EU Member States must take measures to promote low pesticide inputs and better pest management. Member States should also produce a system for the implementation of integrated pest management that ensures farmers have sufficient information, tools for pest monitoring and advisory services on pest management. Here again PA could fulfil these objectives.

The EC **Air Quality Framework Directive from 2008** and **KYOTO protocol** for the reduction of CO_2 emissions aim to reduce the use of fossil fuels. Precision Agriculture and auto-steering systems might help to reduce overlaps and thereby fuel consumption by around 5% (Jensen et al. 2012).

The EC Habitats Directive on the conservation of natural habitats and of wild fauna and flora: Here the adoption of site-specific weed management may improve the natural habitats and wild fauna and flora in relation to reduced use of herbicides.

Weed management is not only targeted to address the needs of the crops but also on economic, environmental and other social aspects in accord with the requirements from the UN Sustainable Development Goals. These are translated into "Good Agricultural Practices" on a global and multinational level by the FAO and EU.

12.4 Stakeholder Involvement

Precision agriculture involves people from a wide range of disciplines including agronomists, remote sensing experts, environmental consultants, agricultural engineers, economists, farmers' advisors, etc. It requires a wealth of knowledge including local and regional farming conditions. All this expertise can rarely be found in a single European country, thus it requires international cooperation to provide the best technology development and decision support tools. A number of stakeholders should be included to improve the adoption of PA systems: policy makers, industry and society, farmers' associations and cooperatives. In addition, public and private advisory services, national, regional and representatives of agricultural authorities (ministries, departments and agencies) have interests related to CAP greening regulations and cross compliance regulations.

Organizations in the field of sustainable agriculture; Research networks and initiatives on precision agriculture and organization, agricultural machinery are also likely to be impacted by the development of the technology. Thus, Commercial exploitation and training of users of PA systems will have to be organised and receive training and extension material developed with plans to improve technology transfer of PA technologies within the following areas:

- Remote sensing applications, including Satellite, Aerial and UAV image applications to help to improve:
- In-field and groundbased detection of nitrogen content in plants
- Groundbased weed detection and weed mapping
- Soil conductivity mapping and (pH) maps to aid lime application
- Weed and fertilizer maps and DSS for fertilizer and pesticide applications

Moreover, precision agriculture and new technology products have to be compliant with the **Environmental Technology Verification programme** of the European Union. In addition, PA should also be in line with the Cross Compliance aspects of the CAP, including Greening, complying with the Statutory Management Requirements as well as maintaining **Good Agricultural and Environmental Conditions** (GAEC).

Social innovation PA will contribute to the digital agenda in the EU by delivering innovative and high-tech services to farmers that will introduce ICT in agriculture, thus contributing to bridging the digital gap in the EU among farmers. This agenda will also help to attract new students to the agricultural colleges and to farm business as such.

Students that have previously been reluctant to enter the farm business may be attracted with a combined interest in both modern technology and traditional farming practices.

12.5 Opportunities for Small and Medium Size Companies

Precision agriculture may also provide market opportunities in the short and medium term given the following:

The current prospects for the PA market to grow are significant in the coming years because of larger farm holdings, professionalization and modernization of technology and technical developments in Information Technology with faster computers for data handling. For instance, Byrne et al. (2013) project a continued rapid pace of advances in semiconductor technology, a key ingredient in the IT evolution, which is an essential part of PA-technology. Furthermore, the rate of decline in prices of microprocessors show no signs of levelling off.

Start-ups and small business companies emerging from precision agriculture migth develop in the short and medium term as a consequence of the dissemination of PA systems. These emerging companies will be closely related to the structure of the agricultural economies of the different countries involved. Local advisors that use PA technologies with agronomic knowledge can provide green IT jobs to young people that are entering the job market.

Local machinery cooperatives will provide PA machinery needed for the application of the N fertilization plan.

In general, precision agriculture technologies might also contribute to building a commercial farm advisory system with combined knowledge on technological, agronomic and economic aspects from various research and development projects throughout Europe. In other parts of the world we see similar developments. In the **US, agricultural policies** have developed in tandem with the CAP through successive farm bills towards a focus on public goods and sustainability issues, although at lower levels than in the EU. The agri-environmental policies include new technologies such as precision agriculture as part of best management practices, Reimer (2015). Public investments and policy reforms will support landscape management practices to be used by farmers and ranchers for sustaining food and ecosystem security. Although U.S. farms have provided increasing supplies of food and other products, they have also been major contributors to global greenhouse gases, loss of biodiversity, natural resource degradation and public health problems, Reganold et al. (2011). Furthermore, it is recognised that to improve sustainability of U.S. agriculture, practices and technologies that address specific production or environmental concerns associated with mainstream conventional farming systems, new approaches are needed that include 2-year crop rotations, precision agriculture with geospatial technologies that describe field variation, classically bred or genetically engineered crops and reduced or no tillage.

12.6 Concluding Remarks

Agriculture is faced with contrasting opposing requirements from the broader society. On the one hand, agriculture needs to expand production to be able to feed a growing global population. Moreover, agriculture is required to produce for several other objectives including energy, textiles, chemicals, and so on. On the other hand, concerns over the environment, climate, biodiversity and other public goods place restrictions on conventional agricultural production. Precision agriculture can be part of the response to these often conflicting issues. Furthermore, the growing demand for higher value food products in terms of human health and quality that require traceability and information about production processes and resource use corresponds with the possibilities offered by precision agriculture technology. The general movement towards greater integration in food supply chains is a natural extension of the requirements for traceability and product information.

Regardless of the societal benefits, the main purpose for investing in new technology is an increase in profitability. For controlled traffic farming and auto-guidance systems, the economic benefits are significant and well documented. However, the promising properties of variable-rate application have so far been realised only by subsidising dissemination of the technology. In Europe, the trend in the reform of the common agricultural policy towards supporting the multifunctionality of agriculture including environmental and climate friendly production provides opportunities for supporting investments in PA technology. Hence, PA's promises of increasing production with reductions in resource use can become available to farmers in Europe.

References

Baligar VC, Fageria NK, He ZL (2001) Nutrient use efficiency in plants. Commun Soil Sci Plant Anal 32:921–950

Basso B, Sartori L, Cammarano D, Grace PR, Fountas S, Sorensen C (2012) Environmental and economic evaluation of N fertilizer rates 1 in a maize crop in Italy: a spatial and temporal analysis. Biosyst Eng 113:103–111

Bindraban PS, Rabbinge R (2012) Megatrends in agriculture – Views for discontinuities in past and furure developments. Glob Food Sec 1:99–105

Byrne DM, Oliner SD, Sichel DE (2013) Is the information technology revolution over? March 27, 2013. Available at SSRN: https://ssrn.com/abstract=2240961 or https://doi.org/10.2139/ssrn.2240961

Dabbene F, Gay P, Tortia C (2014) Traceability issues in food supply chain management: A review. Biosyst Eng 120:65–80

de Fraiture C, Wichelns D (2010) Satisfying future water demands for agriculture. Agric Water Manag 97:502–511

Delgado JA, Del Grosso SJ, Ogle SM (2010) 15546 N Isotopic crop residue cycling studies and modeling suggest that IPCC methodologies to assess residue contributions to N2O-N emissions should be reevaluated. Nutr Cycl Agroecosyst 86:383–390

Doll JP, Orazem F (1978) Production economics, theory with applications, 2nd edn. Wiley, New York

Dubrovsky NM, Burow KR, Clark GM, Gronberg JAM, Hamilton PA, Hitt KJ, Mueller DK, Munn MD, Puckett LJ, Nolan BT, Rupert MG, Short TM, Spahr NE, Sprague LA, Wilbur WG (2010) The quality of our nation's waters – nutrients in the nation's streams and groundwater, 1992–2004. Circular-1350, U.S. Geological Survey

European Commission (2011) Bio-based economy in Europe: state of play and future potential – Part 2. Directorate-General for Research and Innovation, Food, Agriculture & Fisheries, & Biotechnology

European Commission (2013) Overview of CAP reform 2014-2020. Agricultural Policy Perspectives Brief no 5

European Parliament (2014) Precision agriculture: an opportunity for EU farmers – potential support with the CAP 2014-2020. Directorate-General for Internal Policies, Food, Agriculture and Rural Development

FAO (2009) How to feed the world in 2050. Food and Agriculture Organization of the United Nations, Rome

FAOSTAT (2012) www.faostat.fao.org

Gachango FG, Andersen LM, Pedersen SM (2015a) Adoption of voluntary water-pollution reduction technologies and water quality perception among Danish farmers. Agric Water Manag 158:235–244

Gachango FG, Pedersen SM, Kjærgaard C (2015b) Cost-effectiveness analysis of surface flow constructed wetlands (SFCW) for nutrient reduction in drainage discharge from agricultural fields in Denmark. Environ Manag 56(6):1478–1148

Gold MV (2016) Sustainable agriculture: the basics. Ch. 1. In: Etingoff K (ed) Sustainable agriculture and food supply scientific, economic, and policy enhancements. Apple Academic Press, Oakville

González-Dugo V, Zarco-Tejada PJ, Nicolás E et al (2013) Using high resolution UAV thermal imagery to assess the variability in the water status of five fruit tree species within a commercial orchard. Precis Agric. https://doi.org/10.1007/s11119-013-9322-9

Horrigan L, Lawrence RS, Walker P (2002) How sustainable agriculture can address the environmental and human health harms of industrial agriculture. Environ Health Perspect 110(5):445–456

IPCC (Intergovernmental Panel on Climate Change) (2007) Summary for policymakers. In: Solomon S, Qin D, Manning M, Chen Z, Marquis M, Averyt KB, Tignor M, Miller HL (eds) Climate change 2007: The physical science basis. Contribution of working group I to the fourth assessment report of the intergovernmental panel on climate change. Cambridge University Press, Cambridge. https://www.ipcc-wg1.unibe.ch/publications/wg1ar4/ar4-wg1-spm.pdf

Jensen MS, Lind KM, Zobbe H (2009) Enlargement of the European Union and agricultural policy reform. J Eur Integration 31:329–348

Jensen HG, Jacobsen LB, Pedersen SM, Tavella E (2012) Socioeconomic impact of widespread adoption of precision farming and controlled traffic systems in Denmark. Precis Agric 13:661–677

Jepsen MR et al (2015) Transitions in European land-management regimes between 1800 and 2010. Land Use Policy 49:53–64

Lawson LG, Pedersen SM, Sørensen CAG, Pesonen L, Fountas S, Werner A (2011) Four nation survey of farm information management and advanced farming systems: a descriptive analysis of survey responses. Comput Electron Agric 77(1):7–20

Li X, Hu C, Delgado JA, Zhang Y, Ouyang Z (2007) Increased nitrogen use efficiencies as a key mitigation alternative to reduce nitrate leaching in north China Plain. Agric Water Manag 89:137–147

Malthus TR (1798) An essay on the principle of population. J. Johnson, London

Marz N, Warren J (2015) Big Data: Principles and best practices of scalable realtime data systems. Manning Publications Co., Shelter Island

Mosier AR, Bleken MA, Pornipol C, Ellis EC, Freney JR, Howarth RB, Matson PA, Minami K, Naylor R, Weeks KN, Zhu Z (2001) Policy implications of human-acceleratednitrogen cycling. Biogeochemistry 52:281–320

OECD (2016) Farm management practices to foster green growth. OECD Publishing, Paris

Ørum JE, Jacobsen BH (2013) Økonomisk konsekvens ved ændret kvælstofregulering – med udgangspunkt i Limfjorden. Notat til Natur- og Landbrugskommissionen. (Economic assessment of models for regulation of nitrogen leaching – case water catchment area Limfjord). Institute of Food and Resource economics, University of Copenhagen

Pedersen SM (2003) Precision farming – technology assessment of site-specific input application in cereals. Department of Manufacturing Engineering and Management, Technical University of Denmark. Ph.D dissertation, 343 p

Pedersen SM, Pedersen JL (2002) Economic and environmental impact of site specific N-Application – based on different weather conditions and arable crops. In: Robert PC (ed) Proceedings of the 6th international conference on precision agriculture. Minneapolis, pp 1814–1825

Pedersen SM, Pedersen JL, Gylling M (2002) Perspektiverne for præcisionsjordbrug. Danish research institute of food economics, FOI working paper no. 6/2002

Pedersen SM, Boesen MV, Ørum JE (2013) Institutional and structural barriers for implementing on-farm water saving irrigation systems. Food Econ 9(sup5):11–26
Primicerio J, Gennaro SFD, Fiorillo E, Genesio L, Lugato E, Matese A (2012) A flexible unmanned aerial vehicle for precision agriculture. Precis Agric 13:517–523
Rabbinge R, Bindraban PS (2012) Making more food available: promoting sustainable agricultural production. J Integr Agric 22:1–8
Reganold JP et al (2011) Transforming US agriculture. Science 332:670–671
Reichardt M, Jürgens C (2008): Adoption and future perspective of precision farming in Germany: results of several surveys among different agricultural target groups. February 2009, 10(1):73–94
Reimer A (2015) Ecological modernization in US agri-environmental programs: trends in the 2014 Farm Bill. Land Use Policy 47:209–217
Rizoy M (2004) Rural development and welfare implications of CAP reforms. J Policy Model 26:209–222
Robertson GP, Vitousek PM (2009) Nitrogen in agriculture: balancing the cost of 722 an essential resource. Annu Rev Environ Resour 34:97–125
Robertson MJ, Lyle G, Bowden JW (2008) Within-field variability of wheat yield and economic implications for spatially variable nutrient management. Field Crop Res 105:211–220
Scharf PC, Shannon DK, Palm HL, Sudduth KA, Drummond ST, Kitchen NR, Mueller L, Hubbard VC, Oliveira LF (2011) Sensor-based nitrogen applications out-performed producer-chosen rates for corn in on-farm demonstrations. Agron J:1683–1691
Smil V (1999) Nitrogen in crop production: An account of global flows. Glob Biogeochem Cycles 13:647–662
Szakály Z, Szente V, Kövér G, Polereczki Z, Szigeti O (2011) The influence of lifestyle on health behavior and preference for functional foods. Appetite 58(1):406–413
Tscharntke T, Clough Y, Wanger TC, Jackson L, Motzke I, Perfecto I, Vandermeer J, Whitbread A (2012) Global food security, biodiversity conservation and the future of agricultural intensification. Biol Conserv 151(1):53–59
Van der Schans D, Nammen JJ, van der Klooster A, Molenaar K, Krebbers H, Korver R, van Roessel G, Meertens L, Truiman J (2008) Toepassing GPS en GIS in de akkerbouw – Nut en rendement van toepassingen op het gebied van geolandbouw. Praktijkonderzoek Plant & Omgeving B.V., Business-unit AGV, PPO nr. 3250062000
Vanholme B, Desmet T, Ronsse F, Rabaey K, Van Breusegem F, De Mey M, Soetaert W, Boerjan W (2013) Towards a carbon-negative sustainable bio-based economy. Front Plant Sci 4:174
World Bank (2017) Word development indikators database,www.worldbank.org

Abbreviations and Glossary

ADIS Agricultural Data Interchange Syntax (ADIS) is an agricultural-data-format defined in the ISO standard 11787. ADIS is an ASCII 8-bit data-code, in which each number and letter has a defined impact. ADIS is an open (not crypticed) dataformat that can be read by anybody, e.g. a text file. Not all suppliers of agricultural equipment use ADIS.

AE Application efficiency

Aerial photography Remote sensing technique in which a photo of a portion of the earth's surface is taken from an aircraft or satellite in flight.

Algorithm A formula that relates input to output – e.g. sensor input to actuator output. This could for example be the amount of lime according to site-specific soil pH level.

Application map (see Map-Based Variable-Rate Application System)

APSIM Agricultural Production Systems Simulator

ASCII American Standard Code for Information Interchange. A general system that enables transport of data text between two electronic data processing units.

ATV All Terrain Vehicle

Auto-steering Is an automated steering system for tractors either as lightbars og Auto guidance system with RTK.

Beidou Is a Chinese satellite navigation system.

Big-DATA Big data is a term for data sets that are so large or complex that traditional data processing application software is inadequate to deal with them.

CAN Controller Area Network. A "Bit-serial" system for data communication between two different components – developed by the German company BOSCH. Currently, most electronic/hydraulic units in tractors are developed according to the CAN system."CAN is defined by the user. It is not a standard but a defined principle!

CAN-bus Transporting of CAN-signals. A name for a "step by step" described CAN-construction. For example the cable connections in a plug, that is used for connection between a tractor and for example a fertiliser spreader or injection

S.M. Pedersen, K.M. Lind (eds.), *Precision Agriculture: Technology and Economic Perspectives*, Progress in Precision Agriculture,
https://doi.org/10.1007/978-3-319-68715-5

sprayer etc.The DIN-standard: 9684-1, 9684-2, 9684-3, 9684-5 describes the rules, which should be followed by the manufacturers of different agricultural equipment that uses "LBS" CAN-bus system.

Canopy management See crop canopy

CAP Common Agricultural Policy in EU

Carrier The radio frequency signal on which information is encoded and then transmitted.

Carrier-Phase Traching Accurate and sophisticated method of determining position requiring two special receivers, which measure small differences in the radio signal

CBA Cost Benefit Analysis. A methodology to assess private and social costs, benefits and surpluses of implementing new technologies or large projects investments.

Chlorophyll A natural substance in plants that gives the green pigment.

CPO-weed Crop Protection Online-weed, A Danish Decision Support System

Crop canopy Over hanging cover. Eg. A canopy of leaves on a crop/field.

Crop scouting Visual assessment of crop condition including growth stage/maturity, plant vigor, presence of decease, weed infestation and insect infestation.

Cp Crude protein.

CDMA Code division multiple access

CTF Controlled Traffic Farming

DAAC Danish Agricultural Advisory Centre. A national centre for agricultural for advice on crop production, livestock production etc. Owned by the farmers associations. DAAC (In Danish LR: Landbrugets Rådgivningscenter) cooperates with the regional farmers associations and their advisory centres. www.lr.dk

Daisy-model A Danish nitrogen simulation model developed DJF.

DAPS Decision Algorithm for Patch Spraying. A Danish decision support program developed by DJF. Can be linked to Bedriftsløsningen.

DCF Discounted cash flow

DEM Digital Elevation Model

DIAS Danish Institute of Agricultural Science. A national research institute for agricultural production, DIAS is a part of the Danish Ministry of Food, Agriculture and Fisheries. www.agrisci.dk

Differential Correction Correction of a GPS signal to improve its accuracy. The correction is performed using a second stationary GPS receiver positioned at a known location. The second receiver computes the error in the signal by comparing the true distance from satellites to the GPS measured distance.

DGNSS Differential NSS

DGPS Differential Global Positioning System. A method of using GPS which improves the position accuracy through differential correction.

DIN Deutsche Industri Norm. German organisation for industrial standards. Similar to DS (Dansk Standard).

Direct injection Electronic steering system for precise dose (injection) of chemicals on field sprayers (see sprayers with injection system).

Directed Soil Samples Soil samples directed on the field in certain management units. Unlike grid samples which are organised in uniform grids.

DM Dry Matter

DSS Decision Support System. A general term for advanced PC-programs or simple guidelines with decision algorithms for treatment of different data (e.g. soil data, yield data and crop canopy status) to specific actions. For example Kemira Loris and Agrosat from Datalogisk can be regarded as decision support systems for precision farming.

EFF Estonian Farmers Federation

EOC Early Operational Capability

EPPO code EPPO weed species code (Bayer)

EM-38 A commercial method for soil conductivity mapping

EMI Electro Magnetic Induction. EMI is used for soil conductivity mapping

ESR2 satellites Satellites developed by the European Space Agency

EU European Union, The number indicate the number of member states EU-15, EU-28 EU 35

Eurostat Statistical Office of the European Union

FAO Food and Agricultural Organisation, UN institution

FARMSTAR A satellite technology-based service devised and delivered by Airbus Defence and Space since

FASSET Farm ASSEssment Tool. An integrated economic and environmental farm simulation model. The model is developed by Danish Institute of Agricultural Science.

FDR Frequency Domain Reflectometry

FDMA Frequency-division multiple access

FMIS Farm Management Information System

FMMIS Farm Machinery Management Information System

FOC Full Operational Capability

FR Feed ration

Galileo A European satellite positioning system under development (also known as GNSS-1 and GNSS-2).

Geo-referenced data Spatial data that pertains to specific locations on the Earth's surface.

GHG Geenhouse gas

GIS Geographic Information System. A system, usually computer based, for the input, storage, retrieval, analysis and display of geographic data. The GIS database is usually composed of map-like spatial representations called layers. These layers may contain information on a number of attributes including land elevation, land use, land ownership, crop yield and soil nutrient levels. (see Kemira Loris, Datalogisk and Fieldstar).

GMES Global monitoring for environment and security

GMO Gen-Modified Organism.

GPS Global Positioning System. A network of satellites controlled by the US Department of Defence, which are designed to help determine a radio receiver's position in latitude, longitude and altitude. GPS is the most common positioning system for precision farming.

GLONASS Global'naya Navigatsionnaya Sputnikovaya Sistema. The Russian global navigation satellite system.

GNP Gross National Product

GNSS Global Navigation Satellite System (GNSS) a broad term for geographical position system receivers, either by GPS, GLONASS, Galileo or Beidou

Greenstar A precision farming system developed by John Deere. John Deere is one of the Worlds leading companies of farm vehicles and tractors. Greenstar is only available for equipment designed by John Deere. Uses a Can technology, which (currently) isn't using DIN standards.Greenstar is not directly compatible with other systems.

Grid samples Samples which are organised in uniform grids.

Grid Sampling Soil sampling method in which a field is divided into square sections (grids) of several hectares or less. Samples are then taken from each section and analysed.

FC Facilitating Conditions

FC Field capacity

FDR Frequency Domain Reflectometry

FPPP Fast PPP

ICM Integrated Crop Management

IDT Innovation Diffusion Theory

IKONOS A high spatial resolution satellite with a ground resolution of 1 m panchromatic and 4 m multispectral (three visible and one near-infrared). The satellite has the potential of providing data for precision farming.

ICT Information and Communications Technologies

Injection system (see direct injection)

IMI Implement Indicator. An electronic LBS-system component, which automatically identifies the tool, when coupling the equipment to the tractor. The tractor terminal thereby identifies whether it is a crop sprayer or fertiliser spreader which is on the tractor.

IRR Internal Rate of Return

IS Information Systems

ISO The International Standard Organisation. ISO is trying to standardise the German DIN/LBS standard. It will be more comprehensive than the DIN/LBS-standard.

ISOBUS Serial control and communications data network

K (potassium) One of the primary nutrients for arable farming

kPa Kilopascal

LANDSAT LAND SATellite. The name given to a series of US scientific satellites used to study the Earth's surface using remote sensing techniques.

LBS Landwirtschaftliger Bus Systeme. A definition of the standardised system for electronic registration and steering of different machinery applied in agriculture. The system allows for the connection of maximum 16 different LBS-job computers (LBS is described in DIN 9684 2-5).

LBS-terminal "User-unit"- terminal in a LBA CAN-bus system. A LBS compatible terminal follows the guidelines in the DIN-standard.

LCA Life Cycle Assessment

LEPA Low energy pressure application

LESA Low energy spray application

SEGES A Danish National Farm Advisory Center Previously Landbrugets Rådgivningscenter. (in English: Danish Agricultural Advisory Centre (see DAAC).

MESA Mid elevation spray application

MJ Megajoules

Micronutrients Trace elements or minor nutrients – materials needed by plants in very small quantities.

ML Megalitres

MM Motivational Model

MPCU Model of PC Utilization

Multispectral Capable of detecting electromagnetic radiation from several spectral bands simultaneously.

Multispectral Scannner An electromagnetic sensor which collects data in several wavelength bands simultaneously.

NDVI Normalized difference vegetation index is a graphical indicator that can be used to analyze remote sensing measurements

NIR Near Infrared Reflectance. A method with electronic detectors that measure the electromagnetic radiation reflected from a sample irradiated with light of several different wavelengths. It is possible to measure content of protein, starch, fibre and moisture in kernels. See www.vegrains.de

N (Nitrogen) A nutrient critical to plant growth.

Nm Nanometer

NPV Net Present Value.

NRTK Network RTK (se RTK)

NSW New South Wales

NUE Nitrogen Use Efficiency

OECD Organisation for Economic Cooperation and Development.

OSR Oil Seed Rape

Panchromatic Images created from radiation with wavelength. Usually produced in grayscale (black and white).

Pesticides A definition of a group of agricultural chemicals used to protect crops: herbicides, fungicides and insecticides.

PCMCIA Personal Computer Memory Card International Association. A PCMCIA card is a small credit-card size data storage device used by most yield monitors. There are three types, which can be used for storage of data etc.

pH A term used to indicate the degree of acidity or alkalinity. A material that has a pH of 7.0 is neutral. Values above denote alkalinity and below denote acidity.

P (Phosphorus) A non-metallic element, one of the three primary plant nutrients.

Pixel "Picture element", the smallest area or element of an image map.

PL Pesticide Load

Positioning System A general system for identifying and recording, often electronically the location of an object or person, e.g. GPS, GLONASS, GALILEO are global positioning systems. Agrimatic is a local system based on positioning from tramlines.

PPS Precise Positioning System. The GPS service available to the US military that provides users full accuracy with a single mobile receiver. It includes access to the P-code and the removal selection availability effects.

PA Precision Agriculture. A term to describe the management of each crop input on a site-specific basis to reduce waste, increase profits and improve the environment. Sometimes, PA also considers the temporal variation and timing of input application. PA often requires the application of the GPS-system (see also PF)

PF Precision Farming. Managing each crop production input on a site-specific basis to reduce waste, increase profits and improve the environment. Sometimes, PF also considers the temporal variation and timing of input application. PF often requires the application of the GPS-system. (see also PA)

Primary nutrients The three primary crop nutrients. Include nitrogen (N), Phosphorus (P) and Potassium (K)

PTO Power Take-Off

PEOU Perceived Ease of Use

PU Perceived Usefulness

Radiometric system A yield monitoring system that consists of a radioactive source and a sensor. The mass flow rate of a crop through a harvester is determined by the degree to which the crop obstructs the flow of radioactive particles from the source to the sensor.

RGB Red, Green, and Blue, cameras combine the colours red, green and blue to depict the range of colours

Real-Time Correction Correction of a GPS signal by immediately sending the differential correction information to the mobile receiver on-the-go.

RTF Random Traffic Farming as opposite to CTF Controlled Traffic Farming

RTK Real Time Kinematic. Procedure whereby carrier-phase corrections are transmitted in real time from a reference receiver to the user's receiver.

Remote sensing The act of detection and/or identification of an object, series of objects, or landscape without having the sensor in direct contact with the object.

ROBOFARM Project (ICT-AGRI ERA-NET Project "Integrated robotic and software platform as support system for farm level business decisions", funded under the European Union Seventh Framework Programme for Research.

ROI Return of Investment

ROS Robotic Operating System.

SAE Society of Automotive Engineers

Salus System Approach to Land Use Sustainability

Satellite Imaging Processes involved in the formation of an image collected by a satellite-based remote sensing device.
SC Social factors
SCT Social Cognitive Theory
Secondary nutrients The secondary plant inputs include calcium, magnesium and sulfur. Required in small amounts.
Sensor-based variable rate application system A system that adjusts product application rate on-the-go based on information received from real-time sensors. (see also online-application, real time canopy management).
Site-specific yield map A representation of field crop yields collected on-the-go by a harvester equipped with an instantaneous yield monitor. Each location/site in a field is assigned a specific crop yield value.
Soil conductivity mapping A map of the electrical conductivity in the soil layers. A simple and efficient way of analysing the soil clay content etc. Commercial systems are EM-38 and Veris.
Soil testing Analysis of soil samples to determine chemical and physical properties of interest.
Soil texture The physical structure or character of the soil determined by the relative proportions of the soil components (sand, silt and clay) of which it is composed.
Soil type A term used to refer to the combination of primary physical constituents of a soil. For example silty clay loam, fine sandy loam and clay. In Denmark different soil types are classified with numbers e.g. JB1, JB3, JB4 and JB6.
Spatial resolution The size of the smallest object that can be distinguished by a remote sensing device.
SPOT Systeme Pour l'Observation de la Terra.The name given to a series of French scientific satellites used to study the earch's surface using remote sensing techniques.
SAR Synthetic Aperture RadarA system to provide broad area high resolution imagery in all weather and at night. Able to penetrate crop canopy (see www.sandia.gov and airsar.jpl.nasa.gov)
STOA EU Scientific Foresight Study
SFT Smart Farming Technologies
TAM Technology Acceptance Model
TDR Time-Domian Reflectometry. A method to measure soil water content.
TFI Treatment Frequency Index An index used to regulate the application of pesticides in different crop rotations
ToF Time of Flight
TPB Theory of Planned Behaviour
TRA Theory of Reasoned Action
UAS Unmanned Arial System
UAV Unmanned Aerial Vehicle
UGV Unmanned Ground Vehicles
UTAUT Unified Theory of Acceptance and Use of Technology

VRA Variable Rate Application. Adjustments of the amount of cropping inputs such as seed, fertilisers and pesticides to match conditions in a field.

VRT Variable Rate Technology. The equipment used to perform variable-rate applications of crop production inputs.

Vegetation indexes A tool for identifying the levels of health of plant biomass. A vegetation index can be used to assess or predict plant characteristics such as leaf area, total plant material and plant stress. A vegetation index reduces several wavelengths of sensor data into a single number.

Veris A commercial system for soil conductivity mapping.

WARTK Wide Area RTK

Yara N-sensor (previously Hydro N-sensor) A commercial ground based N-sensor designed to measure the variable reflectance of greenness (chlorophyll content) in the crop canopy. The N-sensor is usually mounted on a tractor and measures the crop canopy while distributing fertilisers on-the-run. The system is developed by Hydro Agri.

Yield map A picture of the field, which shows the site-specific yield variability on the field. A yield map is usually made in a GIS-based digital program like for example Fieldstar, Kemira or Agrosat.

Index

S.M. Pedersen, K.M. Lind (eds.), *Precision Agriculture: Technology and Economic Perspectives*, Progress in Precision Agriculture,
https://doi.org/10.1007/978-3-319-68715-5

MIX
Papier aus verantwortungsvollen Quellen
Paper from responsible sources
FSC® C105338

If you have any concerns about our products,
you can contact us on
ProductSafety@springernature.com

In case Publisher is established outside the EU,
the EU authorized representative is:
Springer Nature Customer Service Center GmbH
Europaplatz 3, 69115 Heidelberg, Germany

Printed by Libri Plureos GmbH
in Hamburg, Germany